T0155589

Practical MATLAB

With Modeling, Simulation, and Processing Projects

Irfan Turk

Apress®

Practical MATLAB: With Modeling, Simulation, and Processing Projects

Irfan Turk
Nilufer, Bursa, Turkey

ISBN-13 (pbk): 978-1-4842-5280-2 ISBN-13 (electronic): 978-1-4842-5281-9
https://doi.org/10.1007/978-1-4842-5281-9

Managing Director, Apress Media LLC: Welmoed Spahr
Acquisitions Editor: Steve Anglin
Development Editor: Matthew Moodie
Coordinating Editor: Mark Powers

Cover designed by eStudioCalamar

Cover image by Raw Pixel (www.rawpixel.com)

Distributed to the book trade worldwide by Springer Science+Business Media New York, 233 Spring Street, 6th Floor, New York, NY 10013. Phone 1-800-SPRINGER, fax (201) 348-4505, e-mail orders-ny@springer-sbm.com, or visit www.springeronline.com. Apress Media, LLC is a California LLC and the sole member (owner) is Springer Science + Business Media Finance Inc (SSBM Finance Inc). SSBM Finance Inc is a **Delaware** corporation.

For information on translations, please e-mail editorial@apress.com; for reprint, paperback, or audio rights, please e-mail bookpermissions@springernature.com.

Apress titles may be purchased in bulk for academic, corporate, or promotional use. eBook versions and licenses are also available for most titles. For more information, reference our Print and eBook Bulk Sales web page at http://www.apress.com/bulk-sales.

Any source code or other supplementary material referenced by the author in this book is available to readers on GitHub via the book's product page, located at www.apress.com/9781484252802. For more detailed information, please visit http://www.apress.com/source-code.

Printed on acid-free paper

This book is dedicated to the lovers of MATLAB.

Table of Contents

About the Author

Irfan Turk, PhD is a math and computer programming instructor and has been working in universities, high schools, and educational institutions for nearly 15 years. He concentrated on applied mathematics for his PhD. Dr. Turk finished the computer science track requirements of his master's degree when he was a student at the University of Texas at Arlington. He is the author of *Python Programming: For Engineers and Scientists* and *MATLAB Programming: For Beginners and Professionals*. Dr. Turk's research interests include but are not limited to numerical solutions of differential equations, scientific computing, mathematical modeling, and programming in MATLAB and Python.

About the Technical Reviewer

Karpur Shukla is a research fellow at the Centre for Mathematical Modeling at FLAME University in Pune, India. His current research interests focus on topological quantum computation, nonequilibrium and finite-temperature aspects of topological quantum field theories, and applications of quantum materials effects for reversible computing. He received an MSc in physics from Carnegie Mellon University, with a background in theoretical analysis of materials for spintronics applications as well as Monte Carlo simulations for the renormalization group of finite-temperature spin lattice systems.

Acknowledgments

I would like to mention and thank a few people who helped me in the preparation of this book. I especially thank Steve Anglin, Associate Editorial Director of Apress, who allowed me the honor of authoring this book. I also gratefully thank Mark Powers, Editorial Operations Manager of Apress, who helped me and guided me in bringing this product to life. I also thank Matthew Moodie, Lead Development Editor at Apress, for being a part of this team, and Karpur Shukla, the reviewer of this project. I would like to thank my primary PhD supervisor, Associate Professor Dr. Maksat Ashyraliyev from the Software Engineering Department of Bahcesehir University, for his priceless contributions to my skills. Finally, thanks go to my friend and colleague Ibrahim Emre Celikkale for his recommendations.

Introduction

This book emerged as a result of programming lecture notes, experiences gathered from different computational algorithms, and reading about mathematical models of real-life problems. The purposes of this book are to introduce and teach MATLAB as a programming language and apply the programming concepts to computational models in finance, numerical methods, simulation of randomness, analyzing data with basic statistics, visualization and animation, computational biology, signal processing, image processing, and sound processing. Apart from these illustrations, the book teaches how to create applications with graphical user interfaces (GUIs).

The intended audiences of this project are people who want to learn MATLAB as a programming language, users of MATLAB who want to excel in problem solving, and advanced users of the language looking to prepare applications with GUIs.

The book has two parts. In the first part, the general concepts of the language such as variables, data types, and common built-in functions are explained. Besides these topics, fundamentals of the language such as algorithms, m files, selection structures, loops, and user-defined functions are presented. In the second part of the book, I illustrate solving problems from different areas. In the examples, the algorithmic approach is explained when necessary. Every new item, whether it is a function or a command, is explained so that the reader can understand the subjects and does not miss anything new.

This product can be used as a textbook, or as supplemental material for undergraduate or low-level one-semester graduate courses in colleges or universities. These courses might include but are not limited to computer programming with MATLAB, science and engineering problem solving with MATLAB, scientific computing, and mathematical modeling with MATLAB. The first part of the book should be taught in such courses. The chapters do not depend on each other in the second part of the book. Therefore, topics from the second part can be selected freely depending on the needs of the class. Prior to learning the content of this book, knowledge of mathematics at the precalculus level helps to understand the modeling concepts, so this is recommended as a prerequisite to taking this course. Although a background in any programming language might help to grasp the algorithms used in the examples, it is not a prerequisite. The book is also a practical resource and textbook for individual learners.

It provides 152 illustrative and instructive examples including the solutions along with the codes.

Source code for this book is accessible via the Download Source Code button located at `www.apress.com/9781484252802`.

CHAPTER 1

Introduction to MATLAB

MATLAB is an abbreviation for the expression Matrix Laboratory. It has been widely used in many kinds of applications and fields of study. MATLAB is a high-level language, the reputation of which has been increasing over time. Since its first use in 1970 by Cleve Moler, a famous mathematician and cofounder of MathWorks, Inc. (the owner of MATLAB), it has shown huge advancement and new tools have been added in the new versions released twice a year.

Due to the fact that it has a remarkable number of toolboxes, MATLAB attracts many users from a variety of different areas ranging from engineering to applied sciences. MATLAB has a large number of built-in functions that make the programmer's job easier when it comes to solving problems. Although it is used primarily for technical computing and addresses toolbox-oriented jobs, MATLAB carries a very practical and easy programming language aspect, as well. One of the important goals of this book is to emphasize the programming language aspect of this powerful software.

MATLAB possesses tools that satisfy the programmer's needs in many applications. Even, these days, specific tasks often require specific software. However, MATLAB can suit programmers' demands in most cases.

The MATLAB prompt displays as a double greater sign (>>) in the command window. Due to the trademark and logo usage guidelines of MathWorks, Inc., the ■> symbols will be used together throughout this book to represent the MATLAB prompt.

In this chapter, you will learn the necessary concepts of coding of the language when it comes to solving real-life applications.

MATLAB Environment

When you run MATLAB, the programming frame is opened. The cursor awaits in the command window with the prompt > preceding it. If you run the student version, the prompt is EDU>.

I. Turk, *Practical MATLAB*, https://doi.org/10.1007/978-1-4842-5281-9_1

Let us explore the console of MATLAB, which has a simple appearance. The default window will look like Figure 1-1.

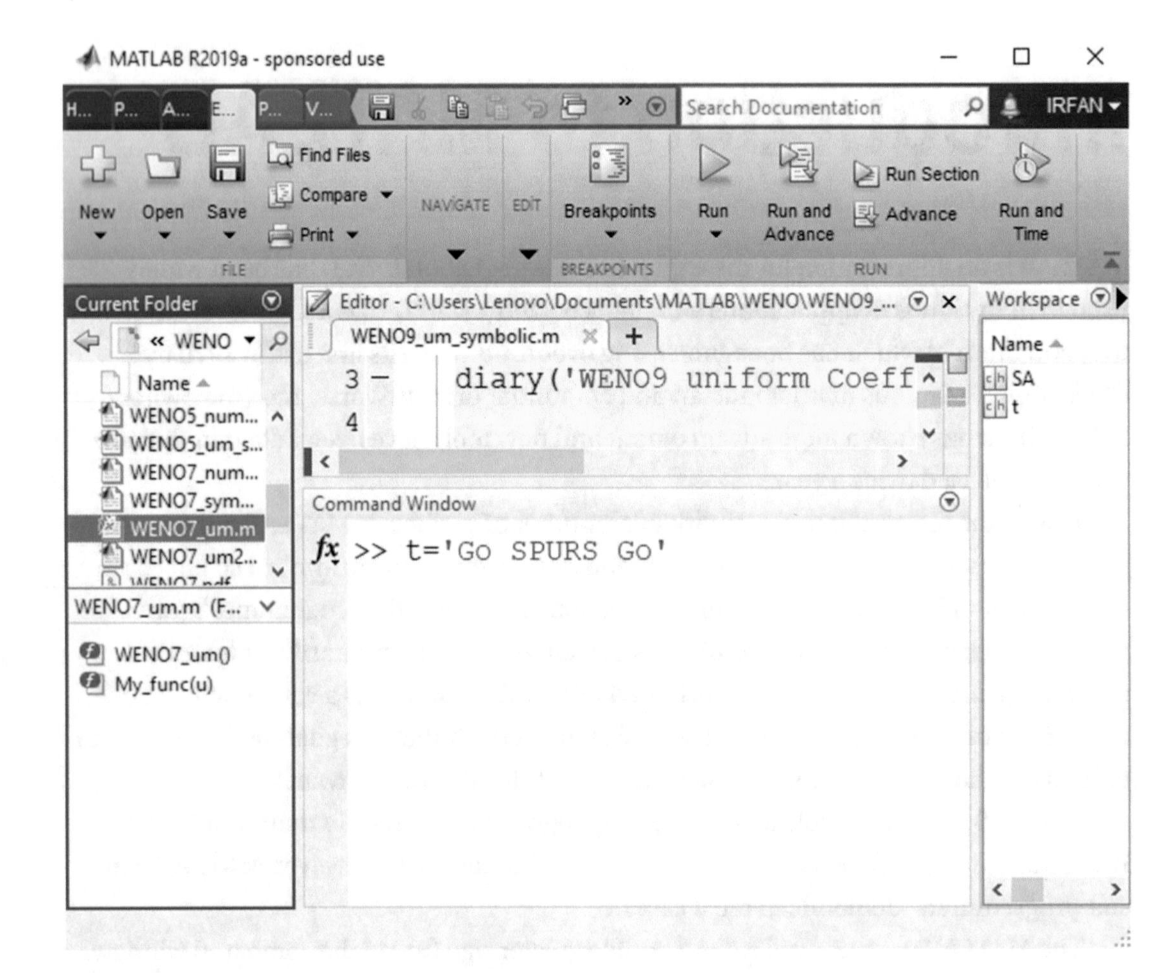

Figure 1-1. *MATLAB environment*

When you open MATLAB, its interface has the following windows:

- *Command Window:* This is the window in which we enter commands.
- *Current Folder:* This window shows the directory in which MATLAB operates.
- *Workspace:* We can see the program variables in this window.

- *Command History:* Here we can monitor the previous commands that we typed in the command window.
- *Editor:* We can write code that we want to run as an m-file in this window.

The user might want to close any of these windows to reorganize the interface. You can also customize the appearance interface, such as the way it looks in terms of color, font, and so on. Click ENVIRONMENT and then Preferencesto open the menu shown in Figure 1-2.

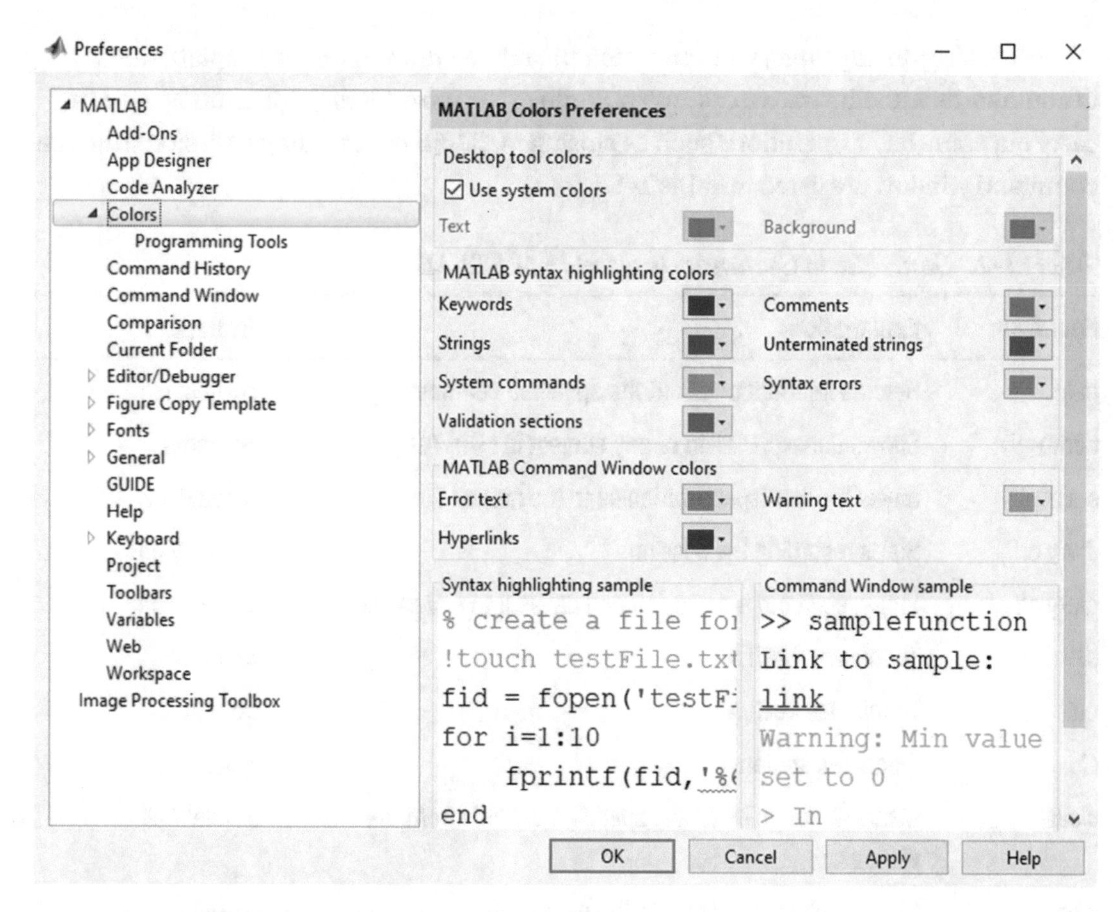

Figure 1-2. *Changing properties*

Just by clicking on the items in the left frame shown in Figure 1-2, you can customize these settings based on your preferences.

When working with MATLAB, one of the most useful commands is the `help` command, which illustrates how a command works and how it is used in MATLAB. Once you type `help` and press Enter, you can click on any of the underlined subjects on the resulting screen to review them in detail.

If you are new to MATLAB, you can watch some tutorials available through a demo. If you type demo in the command window, you can select from any of the available topics to explore:

```
>demo
>
```

By clicking on any topic, you can watch the related tutorials, or see explanations of commands with illustrative examples. Some of the introductory commands used to carry out some basic operations such as closing MATLAB or recording a session from the command window are listed in Table 1-1.

Table 1-1. *Some Basic Commands Used in MATLAB*

Function	Explanation	Example
help	Returns information about the specified command	>>help clc
demo	Shows the explanation of any subject in MATLAB	>>demo
save	Saves the workspace variables to the named file	>>save my_var
diary on	Starts recording the session	>>diary on
diary off	Stops recording the session and saves it to a diary file	>>diary off
exit	Terminates MATLAB	>>exit
quit	Terminates MATLAB	>>quit
clc	Clears the screen	>>clc
clear	Clears all variables or any specified variable from the workspace	>>clear all
who	Displays all the variables in the workspace	>>who
whos	Displays all the variables in the workspace with sizes and types	>>whos

Throughout the book, examples demonstrate the usage of MATLAB. Each example illustrates an important aspect of a feature of the subject in the relevant chapter.

In MATLAB, you can save your session from start to finish by saving your session with the save function. We can take a look at the following example of this.

Example 1-1. Type disp ('Hello World') at the prompt. Save your session in a file named my_session.

Solution 1-1. The following piece of code can be typed at the prompt.

```
> diary my_session
> disp('Hello World')
Hello World
> diary off
```

MATLAB will create a file named my_session in your current directory. If you click on the file, it will be appear in the editor as follows.

```
disp('Hello World')
Hello World
diary off
```

In this code, the command disp() displays whatever is typed between the single quotation marks next to it. If the expression to be displayed is a number, then there is no need to include the quotation marks. If the expression is a string or a letter, then we need to include the quotation marks before and after the text along with the disp() function.

Using MATLAB as a Calculator

MATLAB can be used as a calculator, as well. You can find the solution for any complex calculation. In the following example, we can see an illustration of this function.

Example 1-2. Find the result of $5-\frac{8}{3}+coscos\left(\pi\right)-\frac{10}{e^2}+\sqrt{7}$.

Solution 1-2. The following code will find the solution.

```
> 5-8/3+cos(pi)-10/exp(2)+sqrt(7)
ans =
    2.6257
>
```

As shown in the preceding code, π is represented by word `pi` in MATLAB. For the Euler's number ***e***, we need to type `exp(2)`, and `sqrt(7)` should be used for finding the square root of 7. Because we did not assign a variable to the result, the result is shown as `ans` and it is printed on the screen, where it stands for answer.

Variables and Expressions

In programming languages such as C, C++, and Java, the type of the variable should be specified before the variable is used. However, in MATLAB, this is not the case. The variables are ready to use by just assigning their values. That makes MATLAB more practical for writing shorter and simpler code more quickly.

Some expressions, such as `if`, `for`, or `end` are reserved for the scripts of the language. These words are called *keywords*. To see a list of the keywords used in MATLAB, simply type `iskeyword` at the prompt.

Unlike in Example 1-2, we can assign variable names to the solutions to use them later as well. This is called *assigning*. By typing the following command at the prompt,

```
> my_var = 3
my_var =
     3
>
```

we assign 3 to the variable named `my_var`. Any defined variable is stored as a double precision type in MATLAB by default, unless otherwise specified. Here, the variable `my_var` is a 1 x 1 matrix with type double. We examine data types later in this chapter.

To display the defined variables and the information carried by them in the workspace, the function `whos` can be used.

```
> whos
  Name        Size             Bytes     Class     Attributes
  my_var      1x1              8         double
>
```

There are certain rules for assigning a name to a variable. Variable names cannot be selected in a random manner; they must meet the following requirements:

- They should start with a letter.
- They can contain numbers and underscores.
- They can be a maximum of 63 characters long (the namelengthmax command can be used to check this).
- They should not be a keyword adopted in the MATLAB language.

To avoid confusion, any variable name to be assigned can be checked to see whether it is usable or not at the prompt using the isvarname command.

Another important point that programmers should keep in mind is that MATLAB is a case-sensitive language. In other words, there is a difference between a=5 and A=5 once they are defined in the workspace.

Example 1-3. Check whether it is permissible to use the following names as variable names in MATLAB: Howareyou, hi!, Hola+, Heidi, for, name1, Okay_5

Solution 1-3. We can use the isvarname command to check each of these names. If the result is 1, then it is acceptable to use the name. If the result is 0, the name cannot to be given to a variable. The first two names are checked here as an example:

```
> isvarname Howareyou
ans =
  logical
   1
> isvarname Hi!
ans =
  logical
   0
>
```

Here, 0 or 1 values are assigned to the ans variable, the class type of which is logical.

As you can see, whenever new information is entered at the prompt, it is repeated. If you do not want the computer to repeat what you typed at the prompt, you can insert a semicolon at the end of the line before you press Enter.

Example 1-4. In the equation P∗V = n∗R∗T, the variables are given as P=10, n=2, R=7, and T= ½. Find V according to the given formula.

Solution 1-4. We can enter the following in MATLAB for the solution.

```
> P=10;
> n=2;
> R=7;
> T=1/2;
> V=(n*R*T)/P
V =
     0.7000
>
```

As we can see, there are semicolons after each line except the last one. Because there is no semicolon to the right of the last line, we can see the result of that line after pressing Enter.

Formats

In MATLAB, there are line spacing formats and various numerical formats. Line spacing formats control the spacing between the lines in showing the results at the command window. Numerical formats shape the representation of the output.

Per the line spacing format, there are two options: *compact* and *loose*. The compact option keeps the lines tight and closer, whereas the loose option introduces additional spacing between the lines in the command window.

Example 1-5. Let A=22/7. Show A both in compact format and loose format in the command window.

Solution 1-5. If you type the following in the command window, you will see the variable A in both formats.

```
> A=22/7;
> format compact
> A
A =
     3.1429
> format loose
```

```
> A
A =
    3.1429
>
```

We use the compact format throughout this book to save space.

As for the numerical formats, more options are available. If the format is not altered, the default, *format short,* is used. This format yields calcuations to four decimal places by default. Different alternatives are shown in Table 1-2.

Table 1-2. *Numerical Format Types*

Style	Display	Example
format short	Shows 4 decimal digits (default)	0.3333
format long	Shows 15 decimal digits	0.333333333333333
format shortE	Shows 4 decimal digits in scientific notation	3.3333e-01
format longE	Shows 15 decimal digits in scientific notation	3.333333333333333e-01
format shortG	Same as format short, or format shortE, whichever is more compact	0.33333
format longG	Same as format long, or format longE, whichever is more compact	0.333333333333333
format shortEng	Shows 4 decimal digits in engineering notation	333.3333e-003
format longEng	Shows 12 decimal digits in engineering notation	333.333333333333e-003
format +	Positive/negative/blank	+
format bank	Shows in currency format with 2 digits after decimal points	0.33
format hex	Shows the hexadecimal representation	3fd5555555555555
format rat	Converts the decimal number to a fraction	1/3

Example 1-6. Let A=22/7. Show A in the formats of long scientific notation, short engineering notation, hexadecimal format, and fraction.

Solution 1-6. The commands used in the solution and the corresponding output are shown here.

```
> format longE
> A
A =
     3.142857142857143e+00
> format shortEng
> A
A =
     3.1429e+000
> format hex
> A
A =
   4009249249249249
> format rat
> A
A =
       22/7
>
```

Vectors and Matrices

MATLAB's foundation is based on matrices. In other words, the basic data type in MATLAB is a matrix. There exist close relations among arrays, vectors, and matrices. In this section, we explore arrays, vectors, matrices, and the colon operator used in MATLAB.

Arrays

An *array* is a group of objects having the same type, size, and attributes. An array that consists of one element is called a *scalar.* Arrays can be specified as vectors or matrices. A list of arrays arranged as a column or as a row is a one-dimensional matrix. We can think of a 4×5 matrix, having two dimensions as shown in Figure 1-3.

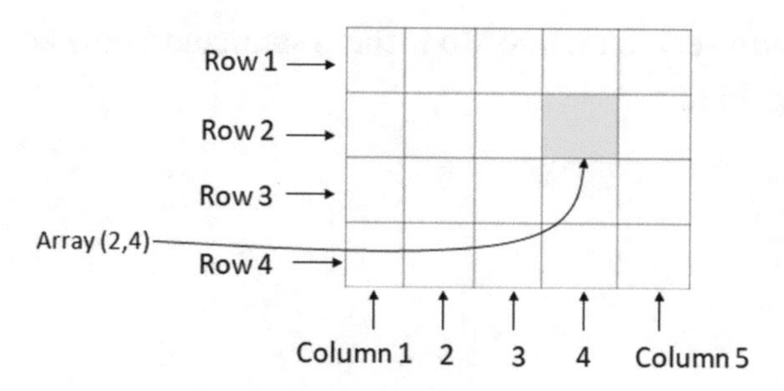

Figure 1-3. *An array*

In Figure 1-3, the cell filled with yellow represents the array that constitutes the second row and the fourth column of the matrix. Therefore, we can think of arrays as elements of matrices.

Vectors

A one-dimensional matrix that represents a row or a column matrix is called a *vector.* In the example shown here, A is a row vector having four elements and B is a column vector having three elements.

$$A = [1 \quad 2 \quad 3 \; 4] \rightarrow \text{1x4 array 4 elements, } \textbf{row vector}$$

$$B = \begin{bmatrix} 1 \\ 3 \\ 5 \end{bmatrix} \qquad \rightarrow \text{3x1 array 3 elements, } \textbf{column vector}$$

$$B(3) = 5;\ A(2) = 2$$

To form a row vector, it is sufficient to leave a space between the cells. Also, a comma can be placed between the numbers as shown at the prompt here.

```
> A=[1,2,3,4]
A =

    1    2    3    4
> A=[1 2 3 4]
A =

    1    2    3    4

>
```

To create a column vector, we need to insert a semicolon between the numbers at the prompt as shown here.

```
> B=[1;2;3]
B =
     1
     2
     3
>
```

Using the size and length commands, you can check the size and length of the vectors A and B just specified.

```
> size(A)
ans =
     1     3
> size(B)
ans =
     3     1
> length(A)
ans =
     3
> length(B)
ans =
     3
>
```

linspace Command

The linspace command provides a very convenient way of forming a vector. Any vector can be created using this command when you want to use elements that are equally spaced.

For example, one could create a vector between 1 and 10 having 10 elements by just typing the following command at the prompt:

```
> Vec=linspace(1,10,10)
Vec =
     1     2     3     4     5     6     7     8     9    10
>
```

The same vector can be obtained in another way, as shown here:

```
>Vec=1:1:10
Vec =
     1     2     3     4     5     6     7     8     9    10
>
```

In this example, the vector starts with 1, and approaches 10 with an increment of 1. That is a very efficient way to create vectors in many problems.

Matrices

An array with more than one dimension is called a *matrix*. As shown here, A is a 3×2 matrix with two dimensions.

$$A = \begin{bmatrix} 1 & 2 \\ 3 & 4 \\ 5 & 6 \end{bmatrix} \quad \rightarrow 3\text{x}2 \text{ matrix } 6 \text{ elements}$$

```
A(3,2) =6
Row #     Column #
```

When creating a matrix in MATLAB, we need to insert semicolons between rows as seen here.

```
> A=[1 2;3 4;5 6]
A =
     1     2
     3     4
     5     6
>
```

Special Matrices

There are a number of special matrices in MATLAB that are often used to make programmers' jobs easier. These matrices can be used instead of needing to create them from scratch, thus saving time in programming. A list of the most useful ones is given in Table 1-3.

Table 1-3. *Special Matrix Functions in MATLAB*

Function	Explanation	Example
eye	Creates an identity matrix	`eye(5)`
ones	Creates a matrix where all the elements are ones	`ones(5)`
zeros	Creates a matrix where all the elements are zeros	`zeros(5)`
diag	Extracts or displays the diagonal part of a matrix	`diag(A)`
sparse	Creates a matrix where all the elements are zeros	`sparse(5,5)`
spdiags	Extracts all diagonals from the matrix	`sparse(A)`
speye	Creates an identity sparse matrix	`speye(5,5)`
rand	Creates a randomly generated matrix with values between 0 and 1	`rand(5)`
magic	Creates magic matrices	`magic(3)`

One of the special matrices from Table 1-3 is shown here.

```
> magic(3)
ans =
      8     1     6
      3     5     7
      4     9     2
>
```

Colon Operator

The colon operator is a very important feature that lets users manipulate matrices and vectors. For vectors, we summarize the operator as shown in Table 1-4.

Table 1-4. *Use of the Colon Operator for Vectors*

Representation	Description	Example

y=a:b	Starts from a, and goes with an increment of 1, up to b	`> y=1:5`
		`y =`
		`1 2 3 4 5`
		`>>`

y=a:step:b	Starts from a, and goes with an increment specified by `step`, up to b	`> y=-10:2:3`
		`y =`
		`-10 -8 -6 -4 -2 0 2`
		`>`

For matrices, we show its usage with rows and columns in Table 1-5.

Table 1-5. *Use of the Colon Operator for Matrices*

Representation	Description	Example

A(: ,k)	Is the kth column of A	`> A=[1 2 3;4 5 6;7 8 9];`
		`> y=A(:,2)`
		`y =`
		`2`
		`5`
		`8`
		`>`

A(n , :)	Is the nth row of A	`> y2=A(1,:)`
		`y2 =`
		`1 2 3`
		`>`

Example 1-7. In a function given by $y = 3^x$, obtain the values for the vector x = [0, 0.5, 1, 1.5, 2]. Write a code to obtain the results in MATLAB.

Solution 1-7. We can create vector x, and calculate the values for the function as shown here.

```
> x=0:0.5:2
x =
           0    0.5000    1.0000    1.5000    2.0000
> y=3.^x
y =
      1.0000    1.7321    3.0000    5.1962    9.0000
>
```

Example 1-8. In a matrix given by $A = [0 - 1\ 2\ 4\ 2\ 3\ 9\ 8\ 5\ ]$, two vectors, B and C, are defined as the second column, and the second and third elements of the third row of matrix A, respectively. Write code to obtain the vectors B and C.

Solution 1-8. Using the colon operator, we can easily obtain B and C as shown in the following prompt.

```
> A=[0 -1 2;4 2 3;9 8 5];
> B=A(:,2)
B =
    -1
     2
     8
> C=A(3,2:3)
C =
     8     5
>
```

Built-in Functions

MATLAB has numerous built-in, ready-to-use functions that make the programmer's job much easier. Although it is hard to categorize these functions precisely, we can group the most frequently used ones into categories such as elementary math functions, trigonometric functions, complex numbers, random numbers, and basic plotting functions, although there are more than we can list here. Using the `help` command, you can easily review the descriptions of the functions and their example uses. For a list of all elementary math functions, simply type `help elfun` at the prompt in MATLAB. Most of the other built-in MATLAB functions are introduced and explained in subsequent chapters.

Some of the Elementary Math Functions

In this section, I present the exponential functions and some other important functions that are used for rounding and finding remainders of a division function.

Table 1-6. *Some Math Functions*

Function	Explanation
exp	Exponential function
log	Natural logarithm function
log10	Common logarithm function in base 10
reallog	Natural logarithm of a real number
sqrt	Square root of a number
nthroot	Real nth root of a real number

Example 1-9. Calculate $y=\frac{8}{\sqrt{27}}$ in MATLAB.

Solution 1-9. If the following code is typed at the prompt, the answer will be obtained as follows.

```
> y=log2(8)/sqrt(27)
y =
    0.5774
>
```

Table 1-7. Additional Math Functions

Function	Explanation
fix	Rounds number toward zero
floor	Rounds number toward minus infinity
ceil	Rounds number toward plus infinity
round	Rounds number toward nearest integer
mod	Shows remainder after dividing
rem	Shows remainder division
sign	Returns -1,0, or 1 (Signum function)

Example 1-10. Find the values of x and y, where $y = \lceil 2.9 \rceil + \lfloor 12.8 \rfloor$ and $x = mod(157,5)$.

Solution 1-10. For the first variable *y*, we are asked to find the sum of the upper integer of 2.9 and lower integer of 12.8. For the second variable *x*, we are supposed to find the remainder of the division of 157 by 5. The code that finds the solution is given here.

```
> y=ceil(2.9)+floor(12.8)
y =
     -9
> x=mod(157,5)
x =
     2
>
```

Trigonometric Functions

Commands for the trigonometric functions in MATLAB are very intuitive. These functions are listed in Tables 1-8 and 1-9.

***Table 1-8.** Trigonometric Functions in Radians*

Command	Definition
sin	Sine
cos	Cosine
tan	Tangent
cot	Cotangent
sec	Secant
csc	Cosecant

***Table 1-9.** Trigonometric Functions in Degrees*

Command	Definition
sind	Sine in degrees
cosd	Cosine in degrees
tand	Tangent in degrees
cotd	Cotangent in degrees
secd	Secant in degrees
cscd	Cosecant in degrees

Data Types

In MATLAB, the default data type is double. This means that anything entered at the prompt is saved as double unless otherwise specified.

We can divide the data types into two major categories in MATLAB: homogeneous data types and heterogeneous data types (Figure 1-4). Homogeneous data types include the same type of data, whereas the heterogeneous data types have mixed or complex data types.

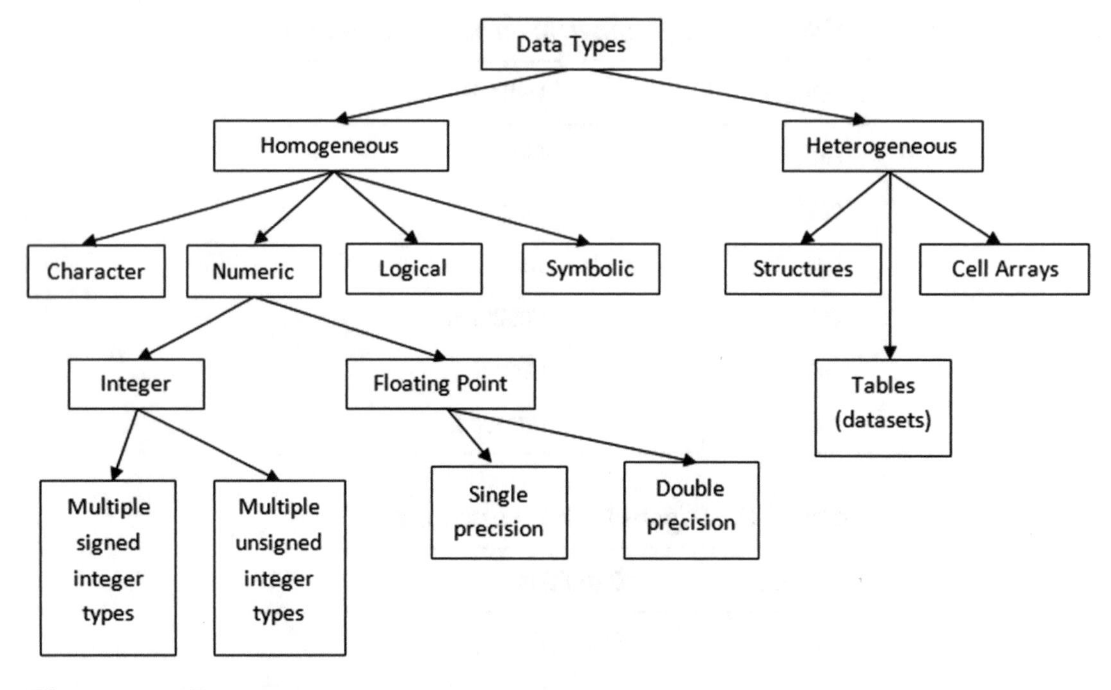

Figure 1-4. *Data types*

Homogeneous Data Types

Homogeneous data types have the same characteristics. These types may be characters, strings, integers, floating-point numbers, or logical data.

Characters and Strings

Characters are composed of strings. Letters, numbers, or symbols can be defined as characters in MATLAB. Characters can be defined by entering them between single quotation marks.

```
> Let='34'
Let =
34
> class(Let)
ans =
char
>
```

In this example, even though we define the variable *Let* as 34 having all numbers, due to the fact that these numbers are entered between the single quotation marks, MATLAB recognizes `Let` as a character. If the `class` command is used to display its class, we will see that it is of type character.

```
> H='How are you?'
H =
How are you?
> length(H)
ans =
    12
> class(H)
ans =
char
>
```

In the preceding example, the variable H is defined as a character of length 12. The variable is defined as a character by putting its name between the single quotation marks.

Example 1-11. Three variables are defined as 'How are you?', 'the weather', and 'is it correct?' for A, B, and C, respectively. We want to create a new character 'How is the weather' by using the given three variables alone. Write the code to achieve this task.

Solution 1-11. To this point, we have typed the commands in the command window. We can write the code in the editor and save it in the working directory of MATLAB. Then by just typing the name of the saved file, we can run the code. This code is written in editor and saved as `CharEx.m`. The following program can be used for this task.

CharEx.m

```
% Name of this file is CharEx.m
% This is Example 1-11
% This code gets some strings and puts them
% together for another string
A = 'How are you?';
B = 'the weather';
C = 'is it correct?';
NewOne = [A(1:4) C(1:3) B]
```

As shown, we extract the first four characters of A, the first three characters of C, and all of the B strings, and assign them into a new variable NewOne. If the percentage (%) symbol is put on a line, MATLAB ignores everything that comes after that % symbol, so this is used for commenting.

Once we run the code at the prompt, the following output will be obtained.

```
> CharEx
NewOne =
How is the weather
>
```

In MATLAB, there is a very useful command called char. Using this command, we can print all the characters defined by the American Standard Code for Information Interchange (ASCII) as shown in Figure 1-5.

ASCII TABLE

Dec	Char	Dec	Char	Dec	Char	Dec	Char	Dec	Char
32		52	4	72	H	92	\	112	p
33	!	53	5	73	I	93	]	113	q
34	"	54	6	74	J	94	^	114	r
35	#	55	7	75	K	95	_	115	s
36	$	56	8	76	L	96	`	116	t
37	%	57	9	77	M	97	a	117	u
38	&	58	:	78	N	98	b	118	v
39	'	59	;	79	O	99	c	119	w
40	(	60	<	80	P	100	d	120	x
41	)	61	=	81	Q	101	e	121	y
42	*	62	>	82	R	102	f	122	z
43	+	63	?	83	S	103	g	123	{
44	,	64	@	84	T	104	h	124	|
45	-	65	A	85	U	105	i	125	}
46	.	66	B	86	V	106	j	126	~
47	/	67	C	87	W	107	k	127	DEL
48	0	68	D	88	X	108	l		
49	1	69	E	89	Y	109	m		
50	2	70	F	90	Z	110	n		
51	3	71	G	91	[	111	o		

Figure 1-5. *ASCII table*

From 32 to 127, these characters can be viewed as it is shown in the solution to Example 1-12.

Example 1-12. Write a program that prints out the lowercase and uppercase letters from the ASCII table.

Solution 1-12. The following code can be used to accomplish the given task.

Letters.m

```
% Name of this file is Letters.m
% This is Example 1-12
% This code prints alphabet letters
Small   = char(97:122);
Capital = char(65:90);
fprintf('Small letters are : %s\n',Small);
fprintf('Big letters are : %s\n',Capital);
```

Then in the command window, once we run the code at the prompt, we obtain the following output:

```
> Letters
Small letters are : abcdefghijklmnopqrstuvwxyz
Big letters are : ABCDEFGHIJKLMNOPQRSTUVWXYZ
>
```

In the code for Letters.m, the `fprintf` function prints purple characters up to the % symbol. After that, the letter `s` tells MATLAB that after that time, strings will be printed that are called `Small` and `Capital`. The `\n` part cuts the line and goes to the next line during printing.

Example 1-13. Bob wants to send the message "Start sending messages at 8:30" to Alice in a secret way as an encrypted message. Write a program that encrypts and decrypts Bob's message. (Hint: Use the ASCII table)

Solution 1-13. The encryption and decryption parts can be given as follows.

HidingMessage.m

```
% Name of this file is HidingMessage.m
% This is Example 1-13
% This code encrypts and decrypts a message
%Encryption starts here
```

```
Message = 'Start sending message at 8:30';
Encryp = double(Message)+3;
fprintf('Encrypted Message : %s\n',char(Encryp))
%Decryption starts here
Decryp = Encryp-3;
fprintf('Decrypted Message : %s\n',char(Decryp))
```

In this solution, the message is converted to the corresponding numbers in the ASCII table using the command `double`. We then add 3 to the corresponding numbers of each letter for encryption. This shifts each letter by 3 to the right in the ASCII table. After the message is encrypted, it is decrypted back to its original version by subtracting 3 from the encrypted values. Once we run the code at the prompt, we obtain the following output:

```
> HidingMessage
Encrypted Message : Vwduw#vhqglqj#phvvdjh#dw#;=63
Decrypted Message : Start sending message at 8:30
>
```

MATLAB provides very useful functions for manipulating strings or characters. Some of the most commonly used functions are listed in Table 1-10.

Table 1-10. *Some of the Built-in Functions Available in MATLAB*

Function	Explanation	Example
fliplr()	Flips the array from left to right	`fliplr('How 5')`
isletter()	States whether the elements are alphabetical letters and returns either 0 or 1	`isletter('trabson61of')`
isspace()	States whether the place is an empty space and returns either 0 or 1	`isspace('Now 12 ')`
lower()	Converts strings to lowercase	`lower('Hola MY friend')`
num2str()	Converts numerical type to string	`num2str('61')`
sort()	Sorts the elements of the array, first the capital letters, and then the small letters	`sort('HOW are you Jack')`

(*continued*)

Table 1-10. *(continued)*

Function	Explanation	Example
str2num()	Converts string to numerical type	`str2num('23')`
strcat()	Adds up the strings horizontally	`strcat('How','Are','You?34')`
strcmp()	Compares the strings; if the strings are the same, it returns 1, else 0	`strcmp('Hola','hola')`
strcmpi()	Compares the strings without case-sensitivity; if the strings are the same, it returns 1, else 0	`strcmpi('Hola','hola')`
strfind()	Finds a string within another string	`strfind('You are','are')`
strncmp()	Compares the first n strings and returns either 0 or 1	`strncmp('ill6','ill7',4)`
strncmpi()	Compares the first n strings without case-sensitivity, and returns either 0 or 1	`strncmpi('ill6','ilL6',4)`
strvcat()	Adds up the strings vertically	`strvcat('How','Are','You?34')`
upper()	Converts strings to uppercase	`upper('Holla My Friend')`

Numerical Data

There exist two different kinds of numerical data types in MATLAB: integers and floating-point numbers. Integers are comprised of signed and unsigned types. There are eight types of integers in total, as shown in Table 1-11. The difference between these types is the storage space that they occupy in memory. If it is possible to do your calculations with low bit integers, you could save space in memory. When higher bit integers are preferred, more memory will be needed.

Table 1-11. Integer Types

Class	Command
Signed 8-bit integer	int8
Signed 16-bit integer	int16
Signed 32-bit integer	int32
Signed 64-bit integer	int64
Unsigned 8-bit integer	uint8
Unsigned 16-bit integer	uint16
Unsigned 32-bit integer	uint32
Unsigned 64-bit integer	uint64

To find the maximum or minimum values of integers within the computer, you just need to type `intmax('classname')`. For example, if we type the following

```
> intmax('int8')
ans =
  127
> intmin('int64')
ans =
 -9223372036854775808
>
```

we see the values of the highest 8-bit integer and the lowest 64-bit integer. Also,

```
> intmin('uint64')
ans =
                    0
> intmax('uint64')
ans =
 18446744073709551615
>
```

shows the lowest and highest unsigned 64-bit integers in the preceding code, using the `intmin` and `intmax` commands.

There are two types of floating-point numbers; namely, double-precision and single-precision numbers. Double-precision floating-point numbers are the default type for all entered variables. Single-precision floating-point numbers occupy 32 bits in memory, whereas double-precision floating-point numbers occupy 64 bits. We can check the highest and lowest values using the same commands we used for integers earlier.

```
> realmax('double')
ans =
  1.7977e+308
> realmin('single')
ans =
  1.1755e-38
>
```

Logical Data

In MATLAB, along with many other programming languages, 1 represents the logical true and 0 represents the logical false.

Example 1-14. Write a program that includes two variables as logical values. One should be true, and the other should be false. Compare these values using the `logical and` and `logical or` operators.

Solution 1-14. The following code can be used to accomplish the given task.

LogicalEx.m

```
% Name of this file is LogicalEx.m
% This is Example 1-14
% This code works with logical values
Logical1 = 3>2 % if it is true, it is 1
Logical2 = 5<4 % if it is false, it is 0
CombineWithAnd = Logical1 && Logical2 % and
CombineWithOr = Logical1 || Logical2  % or
```

If we run the code, we obtain the following result.

```
> LogicalEx
Logical1 =
    1
```

```
Logical2 =
     0
CombineWithAnd =
     0
CombineWithOr =
     1
> class(CombineWithOr)
ans =
logical
>
```

Once we check the class of the variable `CombineWithOr` after running the code, as just shown, it is logical data. Hence, the logical data values can be either 0 or 1.

Symbolic Data

Symbolic data are used to perform algebraic calculations using MATLAB's symbolic toolbox. There are two ways of defining a symbolic variable. One is to type `syms` and the other way is to type `sym` as shown here.

```
> syms x
> class(x)
ans =
    'sym'
> y=sym('y')
y =
y
> class(y)
ans =
    'sym'
>
```

In either method, we can create symbolic data types, and then apply some algebraic operations such as taking the integral or derivative of them.

Heterogeneous Data Types

In some cases, due to the nature of the task at hand, we need to use more complex data types that are obtained by combining more than one type of data. These mixed data types are often called *heterogeneous* data types. Heterogeneous data types are encountered as cell arrays, structure arrays, and dataset arrays (tables) in MATLAB.

Cell Arrays

A *cell* can be regarded as a data box, and a cell array is an array of cells. Using the `cell` function, it is possible to preallocate empty cell arrays. Elements can be indexed by using the cell arrays. As an example, if you type `cell(3,2)` at the prompt, a variable of cell class is created as shown here.

```
> CellAr1=cell(3,2)
CellAr1 =
    []    []
    []    []
    []    []
>
```

To determine its class, you just need to type `class(CellAr1)` for the case given.

```
> class(CellAr1)
ans =
cell
>
```

There are several ways of creating a cell array. You may either preallocate the cell array before assigning values to the array, which is the case shown earlier, or you can just define the array and use it without preallocation.

Example 1-15. Define a cell array that holds the data in Table 1-12.

Table 1-12. *A Cell Array with Six Cells*

1	0	0	'How are you?'	2015
0	1	0		
0	0	1		
'Hola'			72	'Alexander'

Solution 1-15. There are two different ways of defining this cell array. In the first method, all of the elements can be assigned to a single variable, whereas in the second case, the elements should be defined one by one. Let us name the cell array `My_cell_array`. Then, type the following at the prompt;

```
> My_cell_array={eye(3),'How are you?',2015;'Hola',72,'Alexander'}
My_cell_array =
    [3x3 double]    'How are you?'    [       2015]
    'Hola'          [           72]    'Alexander'
>
```

As shown, the index of the cell array is defined within curly brackets.

Structures

The most significant distinction between structures and cell arrays is the indexing. Although cell arrays can be indexed in terms of the elements contained in them, it is not possible to loop through the elements of structures. In other words, data are stored in a *field* in structures.

One of two different methods can be followed to create a structure. One is using the `struct` function and the other one is using the dot operator.

Example 1-16. A company wants to save an employee's information as shown in Table 1-13.

Table 1-13. *Employee Data*

Employee's ID	**5001**
Employee's name	Robert
Employee's address	San Antonio, TX
Employee's salary	39,900

Write a program that stores the information given in Table 1-13.

Solution 1-16. One way of doing that is to create a structure array by using the `struct` function as shown here.

StructArrayEx.m

```
% Name of this file is StructArrayEx.m
% This is Example 1-16
% This code creates a structure
employer=struct('id',5001,'name','Robert','address',...
    'San Antonio, TX','salary',39900)
```

As seen in the preceding code, the data are entered next to one another and are separated by a comma within the `struct` function. Once the program is executed from the command window, the following output will be shown.

```
> StructArrayEx
employer =
  struct with fields:

        id: 5001
      name: 'Robert'
   address: 'San Antonio, TX'
    salary: 39900
>
```

Tables

The third class of heterogeneous data types is called *tables*. Tables are especially convenient for storing column-oriented data. It is possible to perform useful operations on tables such as creating tables, reading data from the tables, changing the content of the tables, and so on. Some basic table functions are shown in Table 1-14.

Table 1-14. *Some Basic Functions Used with Tables*

Function	Explanation	Example
table	Creates tables from the workspace variables	`table(Gender,Smoker)`
readtable	Creates a table from a file	`readtable(filename)`
writetable	Writes a table to a file	`writetable(Table, filename)`
table2cell	Converts a table to a cell array	`table2cell(Table)`
struct2table	Converts a structure to a table	`struct2table(struct)`

Example 1-17. In a workspace, there are three variables:

Name = ['Alex'; 'Slim'; 'Bill'], Age = [35; 40; 45], and Height = [160; 165; 170].

Using these data, write a program to create a table. The table should then be saved in a `MyTable.xlsx` file.

Solution 1-17. The following code can be used to accomplish this task.

CreateTables.m

```
% Name of this file is CreateTables.m
% This is Example 1-17
% This code creates a table and
% saves it as an .xlsx file
Name = ['Alex';'Slim';'Bill'];
Age = [35; 40; 45];
Height = [160; 165; 170];
T = table(Name,Age,Height)
writetable(T,'MyTable.xlsx')
```

Once the code is executed, we obtain the following output.

```
> CreateTables
T =
  3×3 table
    Name     Age    Height
    ____     ___    ______
    Alex     35      160
    Slim     40      165
    Bill     45      170
>
```

The spreadsheet file was saved in the `MyTable.xlsx` file in the directory as well.

More complex applications are available for tables. Here, we have seen only a simple example to give an idea about the general concept.

Plotting Graphics

MATLAB is a very powerful tool for graphics and plotting. A wide range of drawing tools are available for tasks such as plotting in polar coordinates, logarithmic graphics, animated 3-D plots, volume visualization plots, and so on. This section deals with the basic plotting of functions in two dimensions and plotting multiple functions on a single coordinate system or on a single figure.

Single Plotting

The basic command for drawing a function in MATLAB is the `plot` function. The simplest form of the `plot` command is `plot(y)`, where y depends on its index. The most common use of the plot function is in the form of `plot(x,y)`, which is the Cartesian plot of an (x,y) pair.

Example 1-18. Plot the function $y = 2 sinsin\,(x)$ through the interval of $0 \leq x \leq \pi$.

Solution 1-18. If x is defined as a vector from 0 to pi using the `linspace` command, then we can draw the corresponding values of y versus x. In the following code, the distance from 0 to pi is divided into 100 points. If we pick a large number of points such as 1,000, the graph will become more precise.

```
> x=linspace(0,pi,100);
> y=2.*sin(x);
> plot(x,y)
>
```

The code just given produces the output shown in Figure 1-6.

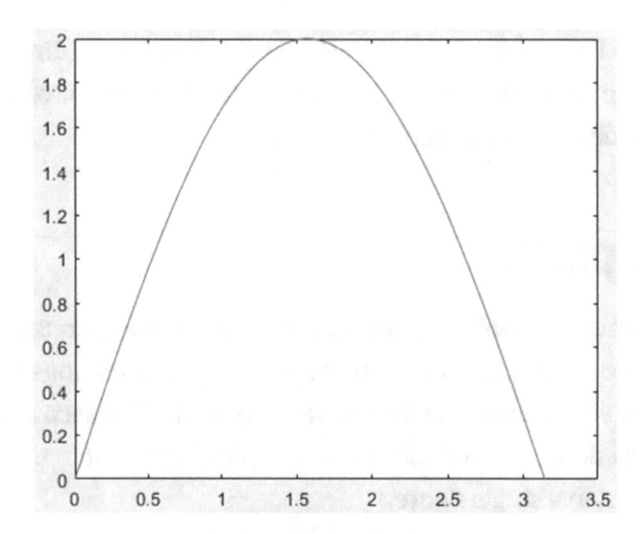

Figure 1-6. *Cartesian plot of y=2sin(x) where* $0 \le x \le \pi$

Various features of graphics such as title, x label, y label, and grids are available in MATLAB. These features are shown in Table 1-15.

Table 1-15. Functions Related to Graphics

Function	Explanation
title('Title')	Adds title to the plot
text(x,y,'string')	Writes string at the point (x,y)
gtext('Text')	Inserts text in the figure manually
xlabel('x')	Prints x horizontally on the plot
ylabel('y')	Prints y vertically on the plot
legend('st1',..,'stN')	Labels each data as st1, … stN strings
grid	Shows the grids on the figure
hold	Keeps the current figure to plot on it
clf	Clears the figure
cla	Clears the axes

There are some other features available that allow programmers to work with line styles, colors, and sizes, as shown in Table 1-16.

Table 1-16. Features of the `plot` Function

Index	Color	Index	Point Type	Index	Line Type
b	Blue	.	Point	-	Solid
g	Green	o	Circle	:	Dotted
r	Red	x	X-mark	-.	Dashdot
c	Cyan	+	Plus	--	Dashed
m	Magenta	*	Star		
y	Yellow	s	Square		

(*continued*)

Table 1-16. (*continued*)

Index	Color	Index	Point Type	Index	Line Type
k	Black	d	Diamond		
w	White	v	Triangle down		
		^	Triangle up		
		<	Triangle left		
		>	Triangle right		
		p	Pentagram		
		h	Hexagram		

To create a regular plot using the `plot` function, different options can be included in the drawing.

There is a remarkable number of special functions available to create two-dimensional and three-dimensional plots in MATLAB. In the command window, if you type `>help specgraph` and press Enter, you will see which special functions are available in your version.

Multiple Plots

It is possible to draw multiple plots on a single figure. That can be achieved in two ways.

We might have multiple plots on the same coordinate system besides having separate plots on one figure. If you want to draw different functions on the same axes, you can do it either by using one single plot function, or using the `hold` command and multiple plot commands.

Example 1-19. Plot the function $y = 2\,sinx$ with grids. Then, keep the first graph and plot a second function given by $y = coscos\,(x)$ within the same interval. Insert the labels for each data set, as well.

Solution 1-19. The desired plot can be obtained using the following program and the results are shown in Figure 1-7.

```
> x=linspace(0,pi,100);
> y=2.*sin(x);
> plot(x,y)
```

```
> hold
Current plot held
> y2=cos(x);
> plot(x,y2)
> title('Title comes here')
> xlabel('This is x label')
> ylabel('This is y label')
> legend('2sin(x)','cos(x)')
> grid on
>
```

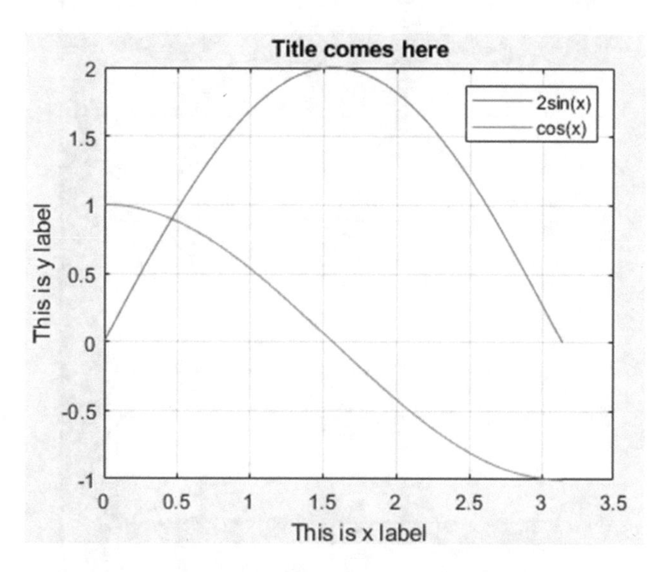

Figure 1-7. *Plots of the functions y=2sin(x) and y2=cos(x)*

It is possible to draw multiple plots using different axes on the same figure via the `subplot` command as well.

Example 1-20. Plot the three functions given in Example 4-5 on different axes on the same figure using the `subplot` command.

Solution 1-20. The following program may be entered at the prompt to accomplish the given task.

```
> x=linspace(0,pi,200);
> subplot(3,1,1)
> plot(x,sin(x))
> subplot(312)
```

```
> plot(x,cos(x))
> subplot(3,1,3)
> plot(x,sin(x)+cos(x))
> grid on
>
```

As shown, `subplot(312)` is used instead of `subplot(3,1,2)`. Both will produce the same result. Once the program is executed, the output shown in Figure 1-8 is obtained.

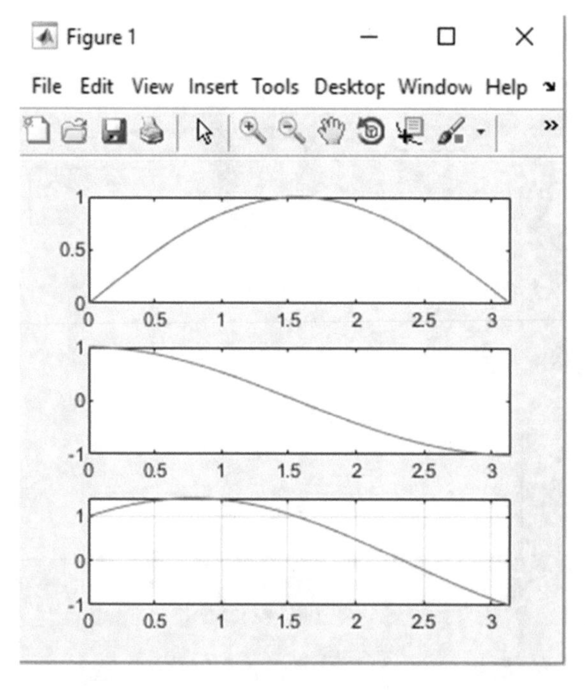

Figure 1-8. *Plots of the three functions on the same figure using* `subplot`

Problems

1.1. Calculate the result for the variable given by $y = \frac{e^3}{14} + 8\sqrt{10}$.

1.2. If you type >>3+6-9 at the prompt and press Enter, to which variable will the result be assigned?

1.3. Which of the following expressions can be a variable name? Check them by using the `isvarname` command.

`Alexander`, `2Hola`, `+Number`, `Good Job`, `How are`, `Bg_3`, `Exam6`, `Add*Or`

1.4. Using the `help` command in the command window, learn the details about the `iskeyword` function. Then, write the difference between the `isvarname` and `iskeyword` commands.

1.5. What is the difference between the expressions `>> disp(4)` and `>> disp('4')` entered in the command window?

1.6. Consider the formula given by F=m*a. Find the value of a when F=45, m=10 is specified.

1.7. If the following is typed at the prompt

```
> Logica=islogical(5<6) && islogical(2>1)
```

what could be your comments about the variable `Logica`?

1.8. Consider the variable, `Numb = 3/8`. Write this variable in the longE, shortG, hex, and rat formats.

1.9. Consider the function $y = cos\,(2x)$, where x =0:pi/12:pi, and obtain the vector y.

1.10. Let $A = [2\ 1 - 3 - 3\ 0\ 5\ 8\ 4\ 50\ ]$. If B is given by the second row of A, and C is given by the intersection of the third column and the first two rows of A, obtain the values of B and C.

1.11. Using Matrix A in Problem 2-2, a new matrix is to be obtained. The first row of A will be removed, and the first and third rows of the resulting matrix will be swapped, as well. What would be the new matrix?

1.12. For $y = \lceil -12.9 \rceil + \lfloor 10.8 \rfloor$ and $x = mod\,(15,9)$, find the values of y and x.

1.13. For $f(x) = \dfrac{2*\left(sin^2x + cos^2x\right)}{tanx}$, calculate the value of $f\left(\dfrac{\pi}{3}\right)$ in radians and degrees.

1.14. Plot the functions $f_1(x) = sinsin\,(2*x)$, $f_2(x) = coscos\,(2*x)$, and $f_3(x) = sinsin\,(2*x) + coscos\,(2*x)$ along the interval $0 \le x \le 2*\pi$ on a single coordinate system.

1.15. Three variables are defined as 'champion Barcelona,' 'is good,' and 'place to go for fun' corresponding to A, B, and C, respectively. You want to create a new statement, 'Barcelona is good place to go,' using the given variables. Write a code to perform this task.

1.16. Write a program that holds the data in Table 1-17.

Table 1-17. *A Cell Array with 12 Cells*

Your	Friend	Is	Ihsan
34	56	52	55
32	20	56	61

Write a program to list the elements of the array, and then display the content of the cell graphically.

CHAPTER 2

Fundamentals of MATLAB Language

In this chapter, you will learn the basic concepts of algorithms, M-files, selection structures, controlling mechanisms of the MATLAB language, and user-defined functions.

A computer program is basically a set of instructions that tells the computer what to do. In MATLAB, for larger programs, sometimes called *scripts,* using an editor is preferred to typing the whole code sequence at the prompt. The problems that we have dealt with so far have been easy to execute and accomplish. It was sufficient to use the prompt alone for solving the problems encountered up to this point. We could also use the MATLAB editor to solve these problems, though. However, for larger and more complex problems, it is recommended that you keep your code together and organized in an editor. For that reason, we use the editor to create our codes for the problems as much as we can for the rest of the book.

It is not difficult to write complex, unclear code. After a while, though, if you look back to your code for some reason, such as upgrading or editing, it might be difficult to understand what you meant. Therefore, one of the primary concerns when creating code is to write clear code that is easy to understand. To accomplish that, we could put our comments within the code as reminders after the % symbol. MATLAB ignores whatever comes after the % symbol. These comments can be placed either after a line of code, or as comment blocks, which are longer informative expressions between %{ and %}.

To generate longer comments, the %{ symbol should be standing alone on a line. Similarly, the %} symbol used for closing the comment block should also be standing alone on the last line, as shown in Figure 2-1.

I. Turk, *Practical MATLAB*, https://doi.org/10.1007/978-1-4842-5281-9_2

```
Untitled3*  x | Example14p12.m  x | Example11p4.m
1    x=2;
2    y ='This is a text';
3    %{
4    Comment starts from here
5    And it ends below.
6    %}
7    z = pi;
```

Figure 2-1. *An example of a comment block*

Another important rule for obtaining more organized codes is to use meaningful names for variables and structures. That helps users recall scripts easily later on. Some other useful tips related to writing better code is outside the scope of this book. For a basic understanding of the subject, though, we keep it simple here.

Algorithms

An algorithm is a computational procedure that shows all the steps involved in solving a problem. In general, an input (or a set of inputs) is (are) taken and an output (or some outputs) is (are) produced. When solving a problem, the problem is separated into parts to analyze and solve the constituent pieces separately. In this way, the problem can be analyzed in a more detailed manner. This method is often called a *top-down* approach. Sometimes, the pieces are combined to get information about the entire system, yielding more complex systems. The system can be constructed by analyzing and combining the pieces about which we have already gathered information. This method is called *bottom-up* processing.

In either method, the purpose of an algorithm is to define a solution to a problem in some way, such as writing the solution via simple expressions, or showing the solution using flowcharts.

Flowcharts and Pseudocode

Both pseudocode and flowcharts are used to construct an algorithm of a computer program. Pseudocode is generally composed of words and expressions, whereas flowcharts are represented by a set of simple shapes.

Pseudocode provides an informal means for describing the steps of the program to be written. It helps the programmer plan and develop the ultimate algorithm for the computer program.

A flowchart is a graphical representation of the steps in writing the program. Simple shapes and diagrams (Table 2-1) are used to describe the algorithm of a computer program.

Table 2-1. *Most Frequently Used Flowchart Symbols*

Symbol	Use
Oval	Used for initiating and finishing the program
Rectangle	Used to indicate calculations, or assigning
Diamond	Used to indicate decision processes
Parallelogram	Used to indicate input or output processes

Example 2-1. Construct pseudocode and a flowchart for an algorithm that calculates the area of a square. The edge length of the square is to be externally entered by the user.

Solution 2-1. For the pseudocode, the steps of the algorithm can be written as follows.

1. Enter the length of the square, called B.
2. Calculate the area of the square, Area= B^2.
3. Display the result, Area.

Scripts and M-Files

After an algorithm is decided on for solving the problem, the code will be generated to accomplish the given task. Code can be written in the MATLAB editor and saved as a file called `*.m`, where * is the name you assign to your program. Thus, the codes written in MATLAB are called m-files and these codes are called scripts or programs.

Example 2-2. Construct the flowchart of an algorithm and generate the code that calculates the area and perimeter of a circle, where the radius is entered by the user.

Solution 2-2. The flowchart of the algorithm can be drawn as shown in Figure 2-2.

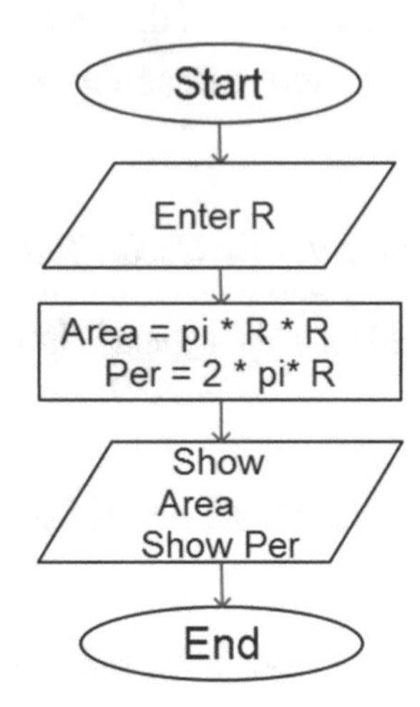

Figure 2-2. *Flowchart of Example 2-2*

Example2p2.m

```
%Example2p2
%this calculates the area and perimeter of a circle
Radius = input('Please enter the radius\n');
AreA = pi*(Radius^2);
Per= 2*pi*Radius;
disp(['Area of the circle is= ',num2str(AreA)]);
fprintf('Perimeter of the circle is = % d\n',Per);
```

Once the code is run, the following output is obtained.

```
> Example2p2
Please enter the radius
10
Area of the circle is= 314.1593
Perimeter of the circle is =  6.283185e+01
>
```

In this solution, four important functions are used. The `input` command is used to externally enter the data. A numerical value is expected to be entered in the preceding code. The code is assigned to the variable named `Radius`. Both the `disp` and `fprintf` functions are used to display data. As seen from the code, though, their usages are different. The `disp` function displays the data between parentheses. After showing the string between single quotation marks, the numerical value of the variable `AreA` is converted to a string by using the `num2str` command, and then it is displayed with the `disp` function. When the `fprintf` function is used, again the data between quotation marks are shown. Here, though, the % symbol tells MATLAB that data will be printed. The letter coming after the % symbol is the specifier of the data. It tells MATLAB what kind of data will be printed. In this example, `d` is used to show a decimal number (Table 2-2). The cursor goes to the next line once `'\n'` is used. As a result, the value of the `Per` variable is printed as a decimal number after the script `Perimeter of the circle is =`.

Table 2-2. *Formatting Text with Conversion Functions*

Specifier	Explanation
c	Displays a single character
d	Displays in signed decimal notation format
e	Displays in exponential notation format
f	Displays in fixed-point notation format
s	Displays character vector or string array

When a conversion function, such as `num2str`, `sprint`, or `fprintf` is used, the user can shape the alignment or precision up to a certain digit. For an illustration, look at the following example.

Example 2-3. Write a program that shows π three different ways. In the first way, the code should display the value of π. In the second case, the code should show three numbers after the decimal. In the third case, the code should print the value of π in 15 digits by aligning π toward the right side.

Solution 2-3. The following code can be used to accomplish the given task.

Example2p3.m

```
%Example2p3
%This code shows three different pi
fprintf('pi :\n %f\n %.3f\n %15f\n',pi,pi,pi)
```

Once the code is run, the following output is obtained.

```
> Example2p3
pi :
 3.141593
 3.142
        3.141593
>
```

In the preceding code, when the cursor comes to `%f\n`, the code prints the value of π and goes to the next line. When it comes to `%.3f\n` for execution, it prints the three numbers after the decimal point and then goes to the next line. Finally, when it comes to execute `%15f\n`, π is shown in 15 digits with the alignment toward the right side padded with spaces as needed. The cursor then goes to the next line.

Logical Functions and Selection Structures

In some situations, we might need to pick an option from among a set of possible candidates. This selection can be achieved by using `if`, `switch`, or `menu` commands in MATLAB. In this section, we discuss each case separately.

if and if-else Commands

There are some cases where it is sufficient to use a single `if` statement alone, whereas the `if-else` structure is used primarily in more complex cases.

Single if Structure

In the case of using a single `if` statement, if the comparison statement after `if` is true, then the commands between `if` and `end` are executed. An example usage of the structure is given here:

```
if comparison
    statement
end
```

Example 2-4. Write code that rolls a pair of dice. If the sum of the two numbers on the top faces is 10, then the computer should print "You are LUCKY." The computer also should print the numbers on the screen.

Solution 2-4. To accomplish this task, the following code, which is saved as Example2p4.m, can be used.

Example2p4.m

```
%Example2p4
% This program rolls a pair of dice
Die=randi(6,1,2);
if Die(1)+Die(2)==10
    fprintf('You are LUCKY \n')
end
fprintf('First # is %d Second # is % d\n',...
Die(1),Die(2))
```

In this code, randi(6,1,2) creates a 1 × 2 matrix randomly where the highest number is 6. The elements are all positive integers in this function, unlike the rand command. The variable Die has two elements. If the sum of these elements is 10, then the message "You are LUCKY" is printed on the screen. In the seventh row of the code, three dots (**...**) are used to tell MATLAB that the coding will continue. Finally, the numbers are printed on the screen.

After running the code, the following output will be obtained.

```
> Example2p4
First # is 1 Second # is  6
>
```

if-else Structure

If using a single `if-else` structure, it is possible to check more than one case in the program. An example usage of this structure is shown here.

```
if comparison
    statement
elseif comparison
    statement
else
  statement
end
```

Example 2-5. Write a program that asks your age. If the entered age is less than 6, MATLAB should print "Maybe NO School" on the screen. If the age is between 6 and 12, it should print "Middle School." If the age is between 13 and 17, it should print "High School," and if the age is between 18 and 25, it should print "Maybe University." Otherwise, the program should display "Professional" on the screen.

Solution 2-5. The following code can be used to accomplish the given task.

Example2p5.m

```
%Example2p5
% This program asks your age
Age = input('Please enter your age\n');
if Age < 6
    fprintf('Maybe NO School \n');
elseif Age <13
    fprintf('Middle School \n');
elseif Age <18
    fprintf('High School \n');
elseif Age <26
    fprintf('Maybe University \n');
else % for other possibilities
    fprintf('Professional \n')
end
```

Once the code is executed, the following output will be shown.

```
> Example2p5
Please enter your age
19
Maybe University
>
```

Relational Operators with if and if-else

Relational operators can be used with `if` and `if-else` structures as well. These operators are listed in Table 2-3.

Table 2-3. *Relational Operators*

Operator	Description	Example
>	Greater than	`3 > 2`
>=	Greater than or equal to	`4 >= 3`
<	Less than	`2 < 3`
<=	Less than or equal to	`2 <= 3`
==	Equal to	`4 == 4`
~=	Not equal to	`2 ~= 3`

For an illustration, let us look at the following example.

Example 2-6. The age intervals that identify your status for getting a driver's license are given in Table 2-4.

Table 2-4. *Ages for Getting a Driver's License*

Interval	License Condition	Comment
Age<16	No license	No license
16<=Age<18	Youth license	Can get a youth license
18<=Age<70	Standard license	Can have a standard license
70<Age	Permitted License	Should get a permitted license

Write a computer program that requests your age and prints one of the comments given in Table 2-4 accordingly.

Solution 2-6. The following code can be used to accomplish the given task.

Example2p6.m

```
%Example2p6
%This program uses relational operators
Age=input('Please Enter Your Age \n');
if Age<16
    fprintf('No license \n');
elseif Age<18 && Age >=16
    fprintf('Can get a youth license  \n');
elseif Age<70 && Age>=18
    fprintf('Can have a standard license \n');
else % for the other possibilities
    fprintf('Should get a permitted license\n');
end
```

If we run the code at the prompt, the output shown here results.

```
> Example2p6
Please Enter Your Age
41
Can have a standard license
>
```

Switch-Case Commands

`Switch-case` commands are very similar to the `if-else` structure. Whatever is programmed using the `if-else` commands can be programmed using the `switch-case` structure as well. There are minor differences between `switch-case` and `if-else` structures. Reading the conditions in a `switch-case` type might be easier compared to the `if-else` type. However, the **switch-case** structure is less flexible than the `if-else` structure due mainly to its nature.

The usage of the switch-case ladder is shown here.

```
switch variable
    case option1
     In case of option1 do these
    case option2
     In case of option2 do these
    otherwise
    If none of the cases is applied, do this
  end
```

Example 2-7. Write a program that, for an amount of money such as $20, $30, $40, or $50 entered by the user, tells you what you can eat from a list including chicken, lobster, beef, and fish, respectively.

Solution 2-7. The following code can be used for the given task.

Example2p7.m

```
%Example2p7
%This program has a switch-case illustration
Money=input('How much money you have? \n');
switch Money
    case 20
        fprintf('You can eat Chicken \n');
    case 30
        fprintf('You can eat Lobster \n');
    case 40
        fprintf('You can eat Beef \n');
    case 50
        fprintf('You can eat Fish \n');
    otherwise % for other possibilities
        fprintf('No match with the menu \n');
end
```

In this code, the number entered should match exactly with one of the numbers used with the case statement. Otherwise, the script written under the otherwise part is printed on the screen. Once the code is run, the following output is obtained.

```
> Example2p7
How much money you have?
40
You can eat Beef
>
```

Menu

The menu function is very useful in listing items. It can either be used with switch-case commands or with if-else commands, which make the appearance nicer. The usage of the menu function is represented as shown here.

```
Name    = menu('memutitle','option1','option2',...,'option')
```

Example 2-8. Write a program that offers four different places to go for vacation using menu. Depending on your choice, the program should tell you how much the vacation costs. The venues include Houston, San Antonio, Dallas, and Austin, and the prices are $450, $550, $650, and $750, respectively.

Solution 2-8: The code generated for this problem is given here.

Example2p8.m

```
%Example2p8
%This code uses menu function
city=menu('Select a city from the menu' ,...
    'Houston','San Antonio','Dallas','Austin');
switch city
    case 1
        fprintf('$450 \n');
    case 2
        fprintf('$550 \n');
    case 3
        fprintf('$650 \n');
    case 4
        fprintf('$750 \n');
end
```

As shown in the first row, three consecutive periods (...) cut the line and extend the command to the next line. Once the program is executed, the interface shown in Figure 2-3 is obtained.

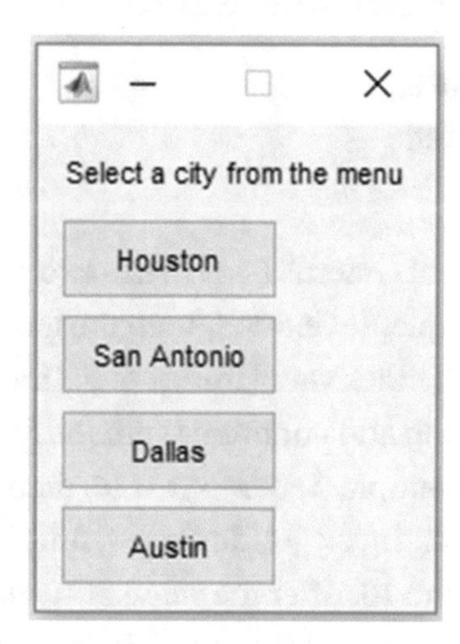

Figure 2-3. *Menu created for the venues*

If the third option, Dallas, is selected, the following output will be obtained.

```
> Example2p8
$650
>
```

Programming Controls

Until a given condition is reached, loop control statements are used to execute a certain part of the code. These loops are sometimes called repetition structures, as well. Similar to most programming languages, MATLAB has `for` loop and `while` loop structures. Within these loops. `break` and `continue` statements can be used to check conditions depending on the cases. Another important control mechanism is to use a `try-catch` block to overcome unexpected situations in MATLAB.

for Loop

If the number of iterations is known, `for` loops are preferred to `while` loops. The configuration of this structure is easy to understand.

```
for starting_variable = Domain
        codes to be executed
end
```

In the given configuration, in general, `Domain` has two or three numbers specified. If it has two numbers, for example, `index=1:5`, then the loop repeats itself five times with an increment of 1 for each index value from 1 to 5. Once the index value equals 6, the computer gets out of the loop and continues with the rest of the script. If `Domain` has three numbers specified, for example, `index =1:3:10`, then the loop repeats itself four times for the values of `index =1`, `index =4`, `index =7`, and `index =10` with an increment of 3 for the index values from 1 to 10. After the value of 10, the program gets out of the loop and continues with the next line of the code. If all indexing operations are not yet complete, then the loop keeps repeating itself. Another important feature of indexing in MATLAB is that indexing starts with 1, and it cannot start with 0.

If a large number is entered as the final value for the index variable, the process of running the code inside the loop could take longer. Depending on your computer's hardware specifications, this might cause your computer to freeze.

The flowchart of a `for` loop is displayed in Figure 2-4.

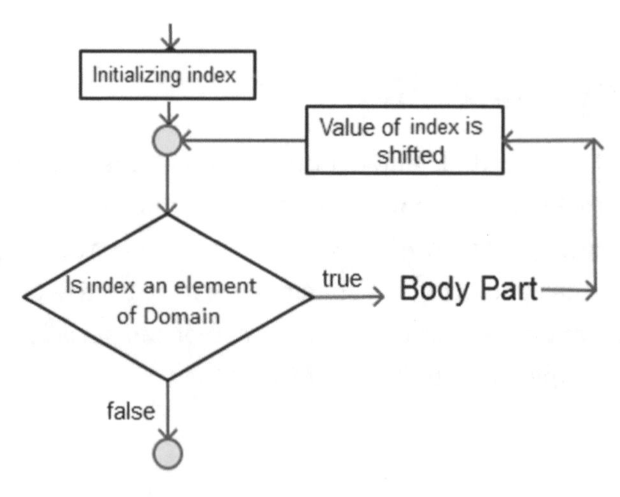

***Figure 2-4.** Flowchart of a for loop*

Example 2-9. Write a program that calculates the following series, which yields the value of ln2.

$$\sum_{i=1}^{\infty}\frac{(-1)^{i+1}}{i}=1-\frac{1}{2}+\frac{1}{3}-\frac{1}{4}+......=ln\,ln(2)=0.693147.$$

Solution 2-9. In programming, a loop cannot be repeated indefinitely in general. Otherwise, it keeps running and does not stop. Therefore, we need to assign a limit to the loop to calculate the series. The higher the limit is, the better the results we obtain. On the other hand, we are limited by another factor, which is the speed of the code. Hence, the code gets slower as the limit chosen is higher. The following code can be used to accomplish the given task.

Example2p9.m

```
%Example2p9
%This code calculates ln2
Total=0;
for i=1:1e+6 %1e+6=1000000
    Number=(-1)^(i+1)/i;
    Total = Total + Number;
end
fprintf('ln2= %d\n',Total);
```

In the beginning, we define a sum equal to zero outside the loop. Then, each iteration is accumulated within that sum. The indexing variable `i` starts from 1, and continues up to 10^6. Once the program is executed at the prompt, the following output will be obtained.

```
> Example2p9
ln2= 6.931467e-01
>
```

As can be seen, the output is shown in scientific form. In other words, the output can be represented by $6,931467.10^{-1}$.

Example 2-10. Write a program that prints a triangle by using asterisks (*). The height is to be entered by the user. For instance, if the entered number is 3, then the code should print the following.

```
>> Height of the triangle? 3
Output:
*
**
***
```

Solution 2-10. We can write the following code.

```
%Example2p10
%This code draws a triangle
x=input('Height of the triangle?\n');
for i=1:x
    for k=1:i
        fprintf('*');
    end
    fprintf('\n');
end
```

The output of this program is shown here.

```
> Example2p10
Height of the triangle?
5
*
**
***
****
*****
>
```

while Loop

In some cases, we might not know the number of iterations within the loop. In such cases, it is preferable to use the `while` loop. Its usage can be summarized in the following form.

```
while condition
        execute
end
```

In this structure, if the condition is true, the program stays in the `while` loop. Unlike the `for` loop, in the `while` loop, we need to add an extra mechanism to control the condition so that we can get out of the loop. In general, one of two options are preferred to get out of a `while` loop. In the first option, the loop is broken using the `break` command if a planned scenario is reached. In the second option, an increment of the indexing variable is defined inside the loop. This is the most significant difference between the `for` loop and the `while` loop.

The flowchart of the `while` loop is almost the same as the one for the `for` loop.

Example 2-11. Write a program that calculates the Euler number within the tolerance at least tol=0.001, where

$$e = \sum_{i=1}^{\infty}\left(1 + \frac{1}{i!}\right)$$

Solution 2-11. The following code can be used for this task.

WhileLo.m

```
%Example2p11
%This code calculates e
Total=1;
Eu=2.718281;
indeks=1;
tol=1; % arbitrarly defined away from Euler
while tol>1e-3
    Number=(1/(factorial(indeks)));
    Total = Total + Number;
    tol=abs(Eu- Total);
    indeks = indeks +1;
end

fprintf('Tolerance: %d e= %d\n',tol,Total);
```

In this program, once the tolerance is not greater than 10^{-3}, the program is terminated. The output becomes this.

```
> Example2p11
Tolerance: 2.254444e-04 e= 2.718056e+00
>
```

break and continue

In some cases, a loop might need to be terminated once a condition is tested. In such cases, the break command is used, which terminates the closest loop and gets out of it.

Example 2-12. Write a program that plays a "Guess the Number" game. The user should guess the number between 1 and 100 where the number is picked randomly by the program. The user will then be guided by the computer to guess either a larger or a smaller number. Finally, the program should print the number that was guessed correctly.

Solution 2-12. The following code can be used.

Example2p12.m

```
%Example2p12
%A Guessing game with break command
Limit=100;%Can be changed
Picked_Number=randi(Limit,1);
Counter=1; %initializing counter
Gues=input('Enter your guess\n');
while true
    if Gues==Picked_Number
        fprintf('Got it in your %d th try\n',Counter);
        break;
    elseif Gues>Picked_Number
        Gues=input('Enter a SMALLER number\n');
    elseif Gues<Picked_Number
        Gues=input('Enter a BIGGER number\n');
    end
    Counter=Counter+1;
end
```

In the code given, a true statement is written next to the while command. In MATLAB, true is a logical variable that always has a value of 1. By writing while true in the seventh row of the program, we want the program to keep running until the selected number is guessed by the user. If the guessed number is equal to the number that the computer picked, then the number of tries is printed and the loop is terminated by the break command.

Once we run the code, the following output will be shown.

```
> Example2p12
Enter your guess
50
Enter a SMALLER number
10
Enter a SMALLER number
5
Enter a SMALLER number
3
Enter a SMALLER number
1
Enter a BIGGER number
2
Got it in your 6 th try
>
```

The `continue` command is used to redirect the code. Unlike with the `break` command, the program continues to work without terminating the closest loop.

Example 2-13. Write a program that asks for the user to enter three positive numbers to calculate its square root. If the entered number is negative, it shows a message indicating that the number should be positive.

Solution 2-13. The following code can be used for this problem.

Example2p13.m

```
%Example2p13
%This code uses continue command
Counter = 3;
while Counter>0
    Number = input('Enter a number\n');
    if Number<0
        fprintf('The number should be positive\n');
        continue;
    end
```

```
    fprintf('Square root of %d is =%d\n',...
        Number,sqrt(Number));
    Counter = Counter-1;
end
```

The following output will be obtained once this code is executed.

```
> Example2p13
Enter a number
-2
The number should be positive
Enter a number
3
Square root of 3 is =1.732051e+00
Enter a number
9
Square root of 9 is =3
Enter a number
4
Square root of 4 is =2
>
```

As shown in the output, if a negative number is entered, the Counter is not decreased, and the program keeps asking for a new number. After entering three numbers, the program stops.

try-catch Block

While running code, unexpected circumstances or errors might occur. To overcome this obstacle and take the necessary actions, the try-catch block is used. This block is composed of two parts. The first part is followed by try and the second part is followed by the catch statement.

```
try
        statement
catch
        statement
end
```

Whatever is written in the `try` statement is regarded as a regular code written in the editor or at the prompt. When an error occurs in this part, the rest of the code within the `try` part is terminated and MATLAB jumps to the `catch` part. This second part is where the user plans to write the alternatives or explanations in the case of getting an error in the first part.

Example 2-14. Write a program that tries to multiply two numbers. The program should call a function for the multiplication that is not defined. The code should give a warning if something goes wrong.

Solution 2-14. The following code can be used to accomplish the given task.

Example2p14.m

```
%Example2p14
%This code uses try-catch block
try
    M1=10;
    M2=20;
    Multp = CallFunctionToMultiply(M1,M2);
    fprintf("Multiplication is %f\n",Multp);
    disp("Everything was smooth")
catch
    warning('Something is wrong');
end
```

Obviously, the program will call the `CallFunctionToMultiply(M1,M2)` function for calculation. User-defined functions are examined in the next section. Here, though, because the function is not defined, the program will give the warning written in the `warning` function. The following output will be obtained once the code shown earlier is executed.

```
> Example2p14
Warning: Something is wrong
> In Example2p14 (line 10)
>
```

User-Defined Functions

Users are allowed to define functions for specific purposes. To define a function in MATLAB, the program must be created within an m-file. In general, user-defined functions are saved as a single m-file to be used for later purposes.

In a user-defined function, the first line should start with the command `function`, generally followed by square brackets including one or more output variables, the function name, and parentheses including one or more input variables. The structure of the user-defined functions can be summarized as follows: In a user-defined function, the command `function` and the name to be assigned are sufficient to create a function. Functions generally end either with `end` statements placed as a final line, or with the definition of another subfunction. Although a user-defined function might occasionally work without the `end` statement, for better readability, using it as the last line of the code is recommended.

```
function [ output_variables ] = function_name( input_variables )
        code to be executed
end
```

As mentioned earlier, the output variables should be between square brackets and the input variables should be enclosed within parentheses. The function name should be identical to the name of the saved file.

Beginning from the second half of 2016, with the R2016b version, another important feature has been incorporated into MATLAB that allows user-defined functions to be added at the end of the code within the same file. This new structure can be summarized as follows.

```
a piece of code
function function1
        code to be executed
end
function function2
        code to be executed
end
```

Creating Functions

When writing a complex program, you might need to create your own functions. Actually, none of the programming languages offers a complete built-in function library for all kinds of problems. Therefore, specific functions can be created for specific purposes to make the programmer's job easier.

Some user-defined functions return a single value, whereas some others return multiple values. A function might not even return any value. In this chapter, we present such types of functions.

Example 2-15. Write a function that prints the square of the number entered by the user. The program should also check whether a number is passed to the function.

Solution 2-15. To create a function, first we need to satisfy the requirements by writing the necessary parts of a function in an editor. We can check the number of inputs by using the `nargin` function to make sure that a number is entered by the user. The following code can be used to do that.

Example2p15.m

```
function [ Result ] = Example2p15( Number )
%Example2p15
%This code accepts ant input
Checking_if_numeric = isnumeric (Number );
if Checking_if_numeric == 0 || nargin ~=1
    error('Just A NUMBER');
end
Result = Number^2;
end
```

The name of the function is the same as the name of the file, `Example2p15`. The function accepts only a single numerical value. This function has an output defined as `Result`, and an input defined as `Number`.

After the program is executed, a logical variable, `Checking_if_numeric`, gets either 0 or 1. If the number is numeric, then 1 is assigned to the logical variable *`Checking_if_numeric`*, 0 otherwise. In the fifth row of the code, the line checks if the variable is zero, meaning that if it is not a number, or the number of inputs is different from 1, then the computer prints the error message, which is placed within the `error` function. In MATLAB, the || symbol is a logical `or` operator.

We will test both error and calculation cases from the command window. If we call the saved function from the command window, and enter a number, such as 5, we obtain the following output.

```
> My_udf(5)
ans =
    25
>
```

If we run the program for a second time, and pass a string such as G to the function, we obtain the following output.

```
> Example2p15(8,4)
Error using Example2p15
Too many input arguments.
>
```

In some cases, we might need to get more than one output from our function as shown in the following example.

Example 2-16. Write a program that calculates the perimeter, area, and volume for a radius entered by the user.

Solution 2-16. The number of outputs should be three. The program is given here.

Example2p16.m

```
function [Perimeter,Area,Volume] = Example2p16(radius)
%Example2p16
%This function may print 3 outputs
Perimeter = 2*pi*radius;
Area = pi*(radius^2);
Volume = (4/3)*pi*(radius^3);
```

If we run the code from command window, we obtain the following result.

```
> [perimeter,area,volume]=Example2p16(10)
perimeter =
   62.8319
```

```
area =
  314.1593
volume =
   4.1888e+03
>
```

As it is seen, there are three outputs specified. If the outputs are not specified, only the first output is displayed.

Local and Global Variables

The variables that are defined inside a function are called *local variables.* Local variables are valid only inside the function, meaning that they are not saved into the workspace. To make these variables available outside the function, we need to add `global` to the left of the variables to define them as global. If you define a function inside another function and use a variable in both functions, again, that variable should be defined as a global variable in both functions. However, many professional programmers recommend avoiding global variables. Instead, passing values to the functions is recommended.

Example 2-17. Write a program that shows the differences between using a global variable inside and outside a function.

Solution 2-17. The following code can be used to accomplish the given task.

Example2p17.m

```
%Example2p17
%This code uses local and global variables.
Numb1 = 61; %this is global variable
fprintf("Variable outside :%i\n", Numb1)
MyFunction();%function is called
fprintf("Variable outside :%i \n", Numb1)
function MyFunction
Numb1 = 55; %this is local variable
fprintf("Variable inside :%i\n", Numb1)
end
```

Once the code is executed, the following output results.

```
> Example2p17
Variable outside :61
Variable inside :55
Variable outside :61
>
```

As shown in the code, the global variable `Numb1` is defined outside the function and printed as 61. Once it is called by the `MyFunction` function, then the same variable is defined as and printed as 55 as well. After exiting the function, its value stays as it was as 61.

Creating Subfunctions

It is possible to create a file that includes all user-defined functions. The first function is defined as the main or primary function. The rest of the functions can be called from the main function to execute the corresponding tasks. Other functions included in the primary function are called subfunctions. To avoid confusion, the name of the primary function and the name of the saved code should be identical.

Example 2-18. Write a function to calculate the volume of a cylinder using a subfunction. The radius and the height of the cylinder should be passed to the function by the user.

Solution 2-18. The code can be written as shown in the following m-file.

Example2p18.m

```
function Example2p18(Radius,height)
%Example2p18
% This PRIMARY FUNCTION calculates
% the volume of a cylinder
radius_sq = R_square(Radius);
Result = pi*height*radius_sq;
fprintf('The volume is %f\n',Result);
end

function [r_sq]=R_square(Radius)
r_sq= Radius^2; %Sub-function
end
```

As shown, the output of the primary function is not defined. Instead, the result is printed on the screen by the body part of the primary function.

The second function used in the program is an example of a subfunction. Once the program is executed, the following output will be shown.

```
> Example2p18(4,10)
The volume is 502.654825
>
```

Anonymous Functions

Anonymous functions are the functions that are not stored in a separate file. These functions are part of a program associated with one or more variables in the file. Anonymous functions possess the `function_handle` data type in MATLAB.

These functions involve the @ symbol in their definitions. The `Sqr` function, given by `Sqr = @(x) sqrt(x)`, is an example of an anonymous function. The @ symbol tells MATLAB that `Sqr` is a function.

Example 2-19. Write a function that takes an input composed of two numbers and calculates their product by using an anonymous function. The product of the numbers should be printed on the screen.

Solution 2-19. The following code can be used for this purpose.

Example2p19.m

```
function [Output] = Example2p19(x,y)
%This is Example2p19
%This function has an anonymous func.
Multip = @(x,y) x*y;
Output =Multip(x,y);
```

After running the code, the following output will be obtained.

```
> Example2p19(5,6)
ans =
    30
>
```

Interaction with the Computer

In this section, we discuss the interaction between the programmer and the computer, in other words, dialog boxes. There are numerous dialog boxes in MATLAB such as errordlg, warndlg, msgbox, questdlg, and so on, to be used within the codes. Some examples of these dialog boxes and their explanations are given in Table 2-5.

Table 2-5. *Examples of Dialog Boxes in MATLAB*

Function	Explanation
inputdlg	Creates an input dialog box
errordlg	Creates an error dialog box
warndlg	Creates a warning dialog box
msgbox	Creates a message dialog box
helpdlg	Creates a help dialog box
waitbar	Opens or updates the wait bar dialog box
questdlg	Creates a question dialog box
listdlg	Creates a list-selection dialog box

To illustrate the usage of a question dialog box, we could write the following code at the prompt.

```
> Picking = questdlg('What kind of Fruit would you like?',...
'Fruit Selection','Apple','Orange','Banana','Orange')
>
```

Once we press Enter, the output shown in Figure 2-5 is obtained.

Figure 2-5. *Output of a question dialog box*

The following syntax appears at the prompt.

```
> Picking = questdlg('What kind of Fruit would you like?',...
'Fruit Selection','Apple','Orange','Banana','Orange')
Picking =
    'Banana'
>
```

To illustrate the use of a list dialog box, the following code can be used.

```
> d= listdlg('PromptString','What kind of Fruit would you like?',...
                'SelectionMode','single', 'ListString',{'Strawberry',...
                'Banana','Orange','Apple','Cherry','Cranberry'},...
                'Name','Select A Fruit','ListSize',[210 100])
d =
     1
>
```

Here, because the first option is picked, as shown in Figure 2-6, 1 is assigned to d.

Figure 2-6. *Output of a list dialog box*

Example 2-20. Write a function to calculate the area of a rectangle. The length and width of the rectangle should be entered by using the input dialog box. The name of the user should be entered into the dialog box, as well.

Solution 2-20. The following code can be used to accomplish the given task.

Example2p20.m

```
%Example2p20
% This example uses inputdlg box
prompt = {'\fontsize{12} Enter your Name:',...
    '\fontsize{12}Length', '\fontsize{12} Width'};
title = 'CALCULATING AREA';
Defining = {'Alex','3','2'};%Default values
NumLines=[1 50]; % dimensions
opts.Interpreter = 'tex';
options.Resize='on';
options.WindowStyle='normal';
Vals=inputdlg(prompt,title,NumLines,Defining,opts);
Name=num2str(Vals{1});%value converted to string
LS=str2double(Vals{2});%value converted to double
SS=str2double(Vals{3});%value converted to double
Area = LS * SS; %Calculation is done
fprintf(" Hello %s\n The Area is :%i\n",Name,Area)
```

The command prompt is a cell array of strings that involve one value for each input. As shown in the preceding code, the font size can be arranged using the `\fontsize{ }` structure. The variable `NumLines` sets the dimension of the inputs. It can be either a scalar value or a matrix. In the preceding code, `str2double` converts a string to double. The command `str2num` can be used, as well.

Once the program is executed, the output shown in Figure 2-7 is the result.

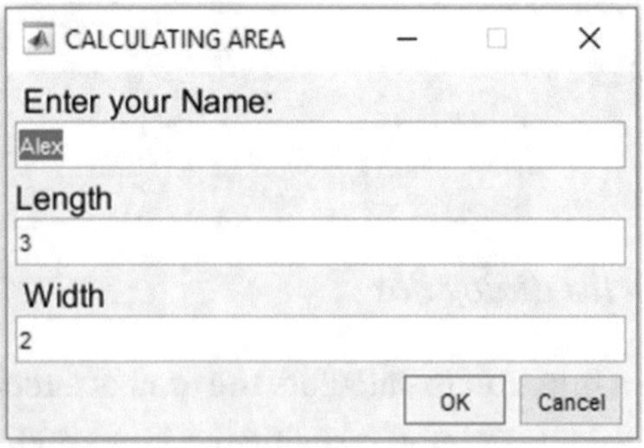

Figure 2-7. *Output of Example 2-20*

After clicking OK without making any changes to the default values, the following output is obtained.

```
> Example2p20
 Hello Alex
 The Area is :6
>
```

Problems

2.1. Construct the code and flowchart of a program that takes two inputs as the length and width of a rectangle, and in turn, calculates the area of the rectangle.

2.2. Construct the flowchart and pseudocode of an algorithm in which, for an entered number greater than 18, the program will print, "Yes you can take a driver's license," and otherwise, "No driver's license."

2.3. Write a program that offers three places to go for a vacation using the `menu` function. Depending on your choice, the program should tell you how much your vacation will cost. The venues include San Antonio, Austin, and Dallas, and the prices are $500, $400, and $300, respectively.

2.4. Write a program that requests the user to pick one of the three colors "Blue," "Red," and "Green," using the `menu` function. The program is required to print the user's choice on the screen.

2.5. Plot the graph of the following piecewise function within the interval of -4≤x≤4.

$$f(x)=\{\frac{x+1}{x^2},-1\leq x\leq 0\,\frac{3-x}{x^2},0\leq x\leq 10,\ |x|>1$$

2.6. Using the asterisk (∗) symbol, write code that prints a rectangle on the screen. The width and the length of the rectangle should be entered as positive integers via the input function.

2.7. Write a program that calculates the Euler number within the tolerance of 0.0000001. The formula is given below to calculate the number.

$$e = \sum_{i=1}^{\infty}\left(1+\frac{1}{i!}\right)$$

2.8. The formula for finding the Fibonacci numbers is $f_n=f_{n-1}+f_{n-2}$, where f(0)=1 and f(1)=1. Write code to calculate f(80).

2.9. Write a function that accepts three inputs. If the number of inputs is not adequate, the program should give an error message.

2.10. Write a program that involves a function in it. The program should calculate the area of a rectangle. The length and width of the rectangle are to be entered after the program is executed. After providing the inputs, these values should be passed to the functions within the program. The function should calculate the area of the rectangle and print that value on the screen.

2.11. Write a program that shows a list of four items. The program should print the name of the selected item from the list on the screen.

CHAPTER 3

Economic Modeling

In this chapter, new functions that are used in the examples are first defined. Following that, simple and compound interest, percentage change, and cost, revenue, and profit topics are examined in different sections. Before the examples are presented, necessary formulas used in the solutions are explained in each section.

Preliminaries

In this section, new terminology used in this chapter is introduced. MATLAB has symbolic data types for calculations in algebraic equations as shown in Chapter 1. The `diff(f,x)` function takes the derivative of symbolic `f` with respect to `x` variable. As an illustration we can take a look at the following input typed at the command prompt.

```
> syms x
> f=x^3-7*x^2+6*x-61;
> y=diff(f,x)
y =
3*x^2 - 14*x + 6
> t=diff(f,x,2)
t =
6*x - 14
>
```

After defining `x` as a symbolic variable and `f`, the derivative of `f` is calculated with respect to `x` with `diff(f,x)` and the calculation is assigned to the `y` variable. Then second derivative of `f` function is calculated with `diff(f,x,2)` and the result is assigned to the `t` variable.

```
> syms x
> y=x^2+2*x-8;
> w=8*x+16;
```


I. Turk, *Practical MATLAB*, https://doi.org/10.1007/978-1-4842-5281-9_3

```
> solve(y,x)
ans =
 -4
  2
> solve(w,x)
ans =
-2
>
```

In the preceding code, using the `solve` function, roots of `y`, and solution of `w` are found that make the equations zero.

```
> syms x
> y=3*x^2-2*x-6;
> MyFunc = matlabFunction(y);
> MyFunc(2)
ans =
     2
> MyFunc(3)
ans =
    15
>
```

Here, using `matlabFunction(y)`, the equation for `y` is converted to a MATLAB function. After that $y(2)$, and $y(3)$ are evaluated with the `MyFunc` function.

Simple and Compound Interest

Interest is the fee paid to borrow money. In most cases, interest is paid when people put their money in bank accounts (i.e., into savings accounts), depending on the contract. In general, two types of interest are calculated, simple and compound interest.

Simple Interest

Simple interest is calculated by multiplying the principal amount, annual rate, and time measured by year. For simple interest, the earned amount is not included in the principal amount at the end of the year. In other words, simple interest is only earned on the principal amount. If the time is not given in years, then it is converted to years. As an example, if the time is given as six months, then time is calculated as $T = \frac{1}{12} * 6 = 0.5$. If it is given as 30 days, then the time is calculated as $T = \frac{1}{365} * 30$ (sssuming that a year is calculated as 365 days). For calculating simple interest, the following formula is used.

$$I = P * R * T$$

where I is the interest, P is the principal, R is the annual interest rate, and T if the time measured in years.

Example 3-1. Write code to calculate payment amount for a certain principal loan, annual rate, and time in months for a bank. The code should ask the user these values, and should print the total payment and interest paid on the screen.

Solution 3-1. The following code can be used to accomplish the given task.

Example3p1.m

```
%Example3p1
%This code calculates simple interest
P = input('Enter Principal Amount: ');
R = input('Enter Annual Rate: ');
T = input('Enter Time As Month: ');
Interest=P*R*(1/12)*T;
Total = P + Interest;
fprintf('Paid Interest Amount is: %.2f\n',Interest)
fprintf('Total Payment is: %.2f\n',Total)
```

In the code, the `input` function is used to enter a value on the screen. Once the value is entered, it is assigned to *P*, *R*, and *T* as specified earlier. Once the code is run, the following output is obtained.

```
> Example3p1
Enter Principal Amount: 2000
Enter Annual Rate: 0.20
Enter Time As Month: 24
Paid Interest Amount is: 800.00
Total Payment is: 2800.00
>
```

Example 3-2. Omar put $80,000 into a bank with a 35% annual interest rate. He plans to withdraw his money at the end of 19 months after depositing it. The bank allows customers to withdraw their funds monthly as well. Write code to calculate the amount of money he will ger at the end of 19 months.

Solution 3-2. The following code can be used to accomplish the given task.

Example3p2.m

```
%Example3p2
%This code calculates interest
P = 8e+4;
R=35/100;
T=(1/12)*18;
I=P*R*T;
Balance = P + I;
fprintf('Balance is %.2f\n',Balance);
```

Once the code is run, the following output is obtained.

```
> Example3p2
Balance is 122000.00
>
```

Compound Interest

Compound interest is different from simple interest. For compound interest, interest is earned on the interest accrued as well. After calculating the interest on principal amount, that value is added to the principal and the new amount becomes the new principal amount. The same process is repeated up to the present or for each cycle. To find only the earned interest, the principal amount is subtracted from the total earnings. For the calculation of compound interest, the following formula is used.

$$I = P*(1+R')^{C} - P$$

where *I* is the Interest, *P* is the principal, *R'* is the periodic interest rate, and C is the compounding period. Periodic interest rate, *R',* is the rate applied to the sum of the principal amount and earned interest for every period. Compounding period, *C,* is the the total number periods for which the interest is applied. Periodic interest rate is calculated by dividing the annual rate by yearly compounding periods.

$$R' = \frac{Annual\ rate}{Periods\ per\ year}$$

As an illustration, for $P = \$100$, annual rate = 15%, periods per year = 3 ($R' = 5\%$), and $C = 4$, total earned amounts (balances) can be calculated for 16 months as shown in Table 3-1.

Table 3-1. *An Illustration*

Period	Interest	Balance
Starting		$100.00
4 months	$100.00*5%=$5.00	$105.00
8 months	$105.00*5%=$5.25	$110.25
12 months	$110.25*5%=$5.51	$115.76
16 months	$115.76*5%=$5.79	$121.55

Example 3-3. Write code to calculate the total balance with a compound interest rate for the information given in Table 3-1.

Solution 3-3. The following code can be used to accomplish the given task.

Example3p3.m

```
%Example3p3
%This code calculates compound interest
format short
P = 100;
Rprime=0.05;
C=4;
Balance = P*((1+Rprime)^C);
disp(Balance)
```

Once the code is run, the following output is obtained.

```
> Example3p3
  121.5506
>
```

Example 3-4. Jennifer puts $150,000 into a bank account that pays a compound interest rate annually of 12%. If she gets her money from the bank after 5 years, how much interest does she get?

Solution 3-4. The following code can be used to accomplish the given task.

Example3p4.m

```
%Example3p4
%This code calculates compound interest
format short
P = 15e+4;
Rprime=0.12;
C=5;
Balance = P*((1+Rprime)^C);
Interest = Balance - P;
fprintf('Obtained interest: %.20f\n',Interest)
```

Once the code is run, the following output is obtained.

```
> Example3p4
Obtained interest: 114351.25248000008286908269
>
```

Percentage Change

We often encounter problems with percentage change in real life. There are three important concepts to understand to solve such types of problems.

- More than x% of $A = A + A * \frac{x}{100} = A * \frac{(100+x)}{100}$
- Less than x% of $A = A - A * \frac{x}{100} = A * \frac{(100-x)}{100}$
- % change $= \frac{new\ value - old\ value}{old\ value} * 100$ percent

Example 3-5. Martin's annual salary is $50,000 this year. He will get a 20% raise for the next year. What will Martin's annual salary be for the next year?

Solution 3-5. The following code can be used to calculate the salary.

Example3p5.m

```
%Example3p5
%This code calculates percentage increment
Salary = 5e+4; %5e+4 = 50000
NextY = Salary*(120/100);
fprintf('Martins Next Year Salary is: %.2f\n',NextY)
```

Once the code is run, the following output is obtained.

```
> Example3p5
Martins Next Year Salary is: 60000.00
>
```

Example 3-6. Martin's market sells melon with a 40% markup at $28 per kilogram. If he wants to sell it with a 20% discount from the original cost, how much does the melon cost Martin per kilogram? Write code that shows the cost and price after discount.

Solution 3-6. The following code can be used to calculate the cost and discounted price.

Example3p6.m

```
%Example3p6
%This code calculates percentage
Melon = 28;
CostAmount = Melon*(100/140);
NewPrc = CostAmount*(80/100);
fprintf('Melon costs: %.2f per kg\n',CostAmount);
fprintf('Price After Discount: %.2f per kg\n',...
    NewPrc);
```

Once the code is run, the following output is obtained.

```
> Example3p6
Melon costs: 20.00 per kg
Price After Discount: 16.00 per kg
>
```

Example 3-7. A bookstore sells a book at $50 and the price was reduced to $35. What is the percentage change on the price of the book?

Solution 3-7. The following code can be used to calculate the percentage change.

Example3p7.m

```
%Example3p7
%This code calculates % change
OldPrice = 50;
NewPrice = 35;
PercChange = ((NewPrice-OldPrice)/OldPrice)*100;
fprintf('Percentage Change= %.2f\n',PercChange);
```

Once the code is run, the following output is obtained.

```
> Example3p7
Percentage Change= -30.00
>
```

As shown in the output, there is a negative (-) sign in front of 30.00. That indicates that the price has decreased. If it is positive (+), then it indicates that there is an increase.

Cost, Revenue, and Profit

In all businesses, keeping a clear and proper financial account is extremely important. Calculation of cost, revenue, and profit plays an important role in this regard. In this section, we deal with these three important concepts.

Cost

Cost, also known as total cost, is the monetary value spent to produce or obtain an item. In general, the cost of something can be calculated by using the linear function

$$C(x) = mx + b$$

The marginal cost is represented by *m*, and *b* represents the fixed costs.

Example 3-8. Alexander's biking company produces bike to sell. Production of each bike costs the company $50 and other fixed costs of the company are $35. Write code to calculate the total cost of produced bikes. The code should ask the for number of bikes produced to complete the calculation.

Solution 3-8. The following code can be used to accomplish the given task.

Example3p8.m

```
%Example3p8
%This code calculates cost (total cost)
Number = input('Number of bikes for production\n');
Cost = @(x) 50*x+35;
CostCalculation = Cost(Number);
disp(['The cost is ', num2str(CostCalculation)])
```

Once the code is run, and number of bikes is entered as 10, the following output is obtained.

```
> Example3p8
Number of bikes for production
10
The cost is 535
>
```

Revenue

Revenue is the total money obtained from selling Q items at price P.

$$R = Q * P$$

In this equation, R represents revenue, Q represents the number of items sold, and P represents the price of each item sold. Finding the total revenue by using the given formula is not that difficult. There are questions that require the user to find the maximum revenue in different cases. The next example is such a question.

Example 3-9. Alexander's cinema company sells tickets for a film showing. The cinema has 1,000 seats. One ticket costs \$8 currently. Alexander wants to increase the price. From past experience, he thinks that if he increases the price \$0.50 for each ticket, then 50 fewer people will attend the showing. Find the price of a ticket that maximizes revenue.

Solution 3-9. Let x be the number of \$0.50 increases. Therefore, \$8.00 + \$0.50x represents the price of one ticket and 1,000 - 50x represents the number of tickets sold. Then the revenue can be calculated as $R = (1000 - 50x) * (8.00 + 0.50x)$. From here, we can write $R = -25x^2 + 100x + 8000$. The following code can be used to accomplish the given task. Then we need to find an x value that maximizes the R value. This is the part where we need to enter into the code and solve for x to make R the maximum value. The following code can be used to achieve this.

Example3p9.m

```
%Example3p9
%This code finds maximum value
syms x
R = -25*x^2+100*x+8000;
Derivat = diff(R,x);
Number = solve(Derivat,x);
EvalFunc = matlabFunction(R);
MaxVal = EvalFunc(double(Number));
disp(['Maximum Revenue is ',...
    num2str(MaxVal)])
```

In this code, the R function is entered as a symbolic function. Then its derivative is found by using the `diff` function. The obtained linear equation is solved by using the `solve` function. The obtained `Number` value shown is a symbolic data value. To convert it to a floating number, the `double` function is used. In the meantime, the R symbolic function is converted to a MATLAB function to evaluate R at `x=Number` value. Finally, the maximum value is shown with the `disp` function.

Once the code is run, the following output is obtained.

```
> Example3p9
Maximum Revenue is 8100
>
```

Remark 3-1. For quadratic equations (i.e., $f(x) = ax^2 + bx + c$, $a \neq 0$ and $a, b, c \in R$), maximum value or minimum value can be calculated depending on the sign of a. If the sign is negative (-), then the maximum value can be derived from the equation. If the sign is positive (+), then the minimum value of the equation can be found. This can be done in two ways. In the first way, $f\left(\frac{-b}{2a}\right)$ gives us the maximum or minimum value. In the second case, the derivative of $f(x)$ is calculated, and the $f'(x) = 0$ equation is solved. The root of $f(x)$ gives us the maximum or minimum value depending on the sign of a as well.

In the preceding problem, the second method is preferred for finding the maximum value.

Profit

Profit can be defined as the difference between the revenue (total revenue) and the cost (total cost).

$$\mathrm{Pr} = \mathrm{R} - \mathrm{C}$$

where *Pr* stands for profit, *R* stands for revenue, and *C* stands for cost.

Example 3-10. A company sells t-shirts with cost function $C(x) = 3 + 2x$, and the revenue function is $R(x) = -x^2 + 4x + 12$ where x indicates the number of t-shirts. For the maximum revenue value *R*, find *x*. Then by using the same *x*, calculate cost (*C*) and profit (*Pr*).

Solution 3-10. The following code can be used for the solution.

Example3p10.m

```
%Example3p10
%This code profit
syms x
R = -x^2+4*x+12;
Derivat = diff(R,x);
Number = double(solve(Derivat,x));
C = @(x) 3+2*x;
Rev = @(x) -x^2+4*x+12;
disp(['For maximum R, x =',...
    num2str(Number)])
Profit=Rev(Number)-C(Number);
disp(['Profit, Pr=',...
    num2str(Profit)])
```

Here, R, which stands for revenue, is defined in the fourth row. Using this, the maximum x variable is calculated by taking its derivative and setting it equal it to zero so that its solution can be found using the `solve` function. Then the same revenue and cost function are defined as anonymous functions. These functions are evaluated at `x=Number`, and the results are shown with the `disp` function. Once the code is run, the following output is obtained.

```
> Example3p10
For maximum R, x =2
Profit, Pr=9
>
```

Problems

3.1. Bushra put her $180,000 into a bank with a 25% annual interest rate (simple). She plans to withdraw her money at the end of 34 months after depositing. The bank allows customers to withdraw their funds monthly as well. Write code to calculate the amount of money Bushra gets at the end of 34 months.

3.2. Alexander puts $240,000 into a bank account that pays a compound interest rate of 20% annually. If he gets his money from the bank after 8 years, how much interest does he get?

3.3. David's market sells watermelon with a 30% markup at $39 per kilogram. If he wants to sell it with a 10% discount from of the original cost, how much does the watermelon cost David per kilogram? Write code that shows the cost and price after discount.

3.4. A coffee shop sells a mug for $20 and then changes the price to $25. What is the percentage change on the price?

3.5. Benjamin's biking company produces bikes to sell. The production of each bike costs the company $40 and other fixed costs to the company are $25. Write code to calculate the total cost of the bikes produced. The code should also calculate the cost of producing seven bikes.

3.6. Lily's cinema company sells tickets for a film showing. The cinema has 1,200 seats. One ticket currently costs $10. Lily wants to increase the price. From past experience, Lily thinks that if she increases the price $0.50 for each ticket, then 20 fewer people will attend the showing. Find the price of the ticket that maximizes revenue.

3.7. For a given function $R(x) = 2x^2 - 4x + 2$ find the minimum value of $R(x)$. Explain the possibility of finding the maximum value of $R(x)$.

CHAPTER 4

Numerical Methods

In this chapter, I present how to solve problems using numerical techniques. First, interpolation and extrapolation methods are introduced. Next, curve fitting and root finding are explained. Finally, numerical integration and numerical differentiation are covered.

Interpolation and Extrapolation

It is possible to find the corresponding value of a point within a given domain using the function `interp1` when dealing with interpolation in one-dimensional applications. It is also possible to find a value that is out of a given domain via the same function. This method is called extrapolation. Various methods such as `linear` or `spline` are available in the `interp1` function (Table 4-1). Some of the methods, which can be used for interpolation, are not available for extrapolation within the `interp1` function. If an outer point is picked for extrapolation using a method in which extrapolation does not work, MATLAB returns NaN which stands for "not a number." Therefore, the method of choice is important for performing interpolation or extrapolation.

Table 4-1. *Methods Used with* `interp1`

Option	Explanation
linear	Uses linear interpolation (default value)
nearest	Uses nearest neighbor interpolation
next	Uses next neighbor interpolation
previous	Uses previous neighbor interpolation
spline	Uses piecewise cubic spline interpolation and extrapolation

I. Turk, *Practical MATLAB*, https://doi.org/10.1007/978-1-4842-5281-9_4

Example 4-1. For the given values of $x = [1, 2, 5, 8]$ and $y = [20, 30, 50, 80]$, calculate the value of 7 using the default method of the `interp1` function, and calculate the value of 10 with the methods `linear` and `spline`.

Solution 4-1. The following code can be used to accomplish the given task.

Example4p1.m

```
%Example4p1
%This code uses interp1 function
x=[1, 2, 5, 8];
y=[20, 30, 50, 80];
ValX1 = interp1(x,y,7);
ValX2 = interp1(x,y,10,"linear");
ValX3 = interp1(x,y,10,"spline");
fprintf("1st value is %f\n",ValX1)
fprintf("2nd value is %f\n",ValX2)
fprintf("3rd value is %f\n",ValX3)
```

Once the code is executed, the following output will be obtained.

```
> Example4p1
1st value is 70.000000
2nd value is NaN
3rd value is 121.428571
>
```

As shown in the preceding output, the computer prints `NaN` for the `2nd value`. It indicates the fact that extrapolation is not available with the `linear` method in the `interp1` function.

Curve Fitting

Using a data set, one can obtain a polynomial function using the function `polyfit`.

Example 4-2. For the given values of $x = [4, 6, 6.5, 8, 10]$ and $y = [4, 8, 6, 7, 9]$, obtain four different curves from the first degree to the fourth. Then, with these four polynomials, evaluate the values of x, where $x = 3{:}0.5{:}20$. The code should plot all the curves in a single figure.

Solution 4-2. The following code can be used to accomplish the given task.

Example4p2.m

```
%Example4p2
%This code uses polyfit function
x=[4, 6, 6.5, 8, 10];y=[4, 8, 6, 7, 9];
P1 = polyfit(x,y,1)%1st degree
P2 = polyfit(x,y,2)%2nd degree
P3 = polyfit(x,y,3)%3rd degree
P4 = polyfit(x,y,4)%4th degree
%Using obtained curves
xx = 3:0.5:20;
y1 = polyval(P1,xx);
y2 = polyval(P2,xx);
y3 = polyval(P3,xx);
y4 = polyval(P4,xx);
subplot(221)
plot(x,y,'s',xx,y1);title('1st degree')
subplot(222)
plot(x,y,'s',xx,y2);title('2nd degree')
subplot(223)
plot(x,y,'s',xx,y3);title('3rd degree')
subplot(224)
plot(x,y,'s',xx,y4);title('4th degree')
```

Once the code is executed, the following output will be printed on the screen.

```
> Example4p2
P1 =
    0.7129    1.8812
P2 =
   -0.0678    1.6691   -1.2124
P3 =
    0.1291   -2.7559   19.2697  -37.1885
P4 =
   -0.2905    8.3000  -86.1202  383.4500 -608.7143
>
```

As shown in this output, the coefficients of each polynomial are calculated and assigned to the corresponding variables. Figure 4-1 will also be displayed along with the output.

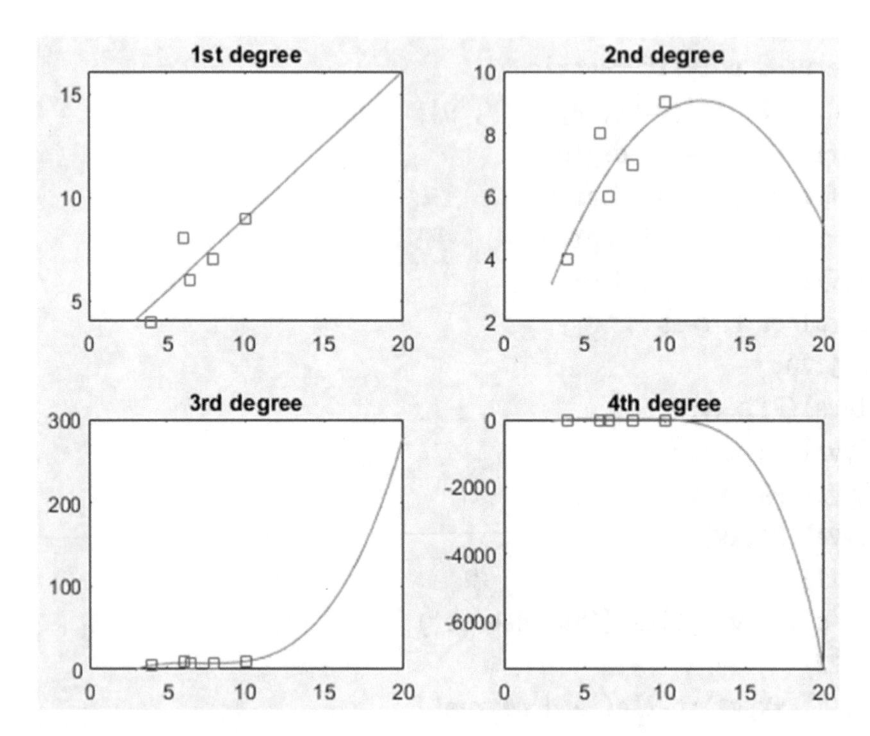

Figure 4-1. *Output of Example4p2*

Root Finding

It is possible to find the roots of nonlinear equations with MATLAB. For this purpose, the built-in function `fzero` can be used. Other popular methods, which we overview in this section, include the bisection method, Newton's method, the secant method, and the fixed-point iteration method.

fzero Function

`fzero` is a built-in function that finds the roots of a nonlinear function in MATLAB. This function is very easy to code and it has two different uses. The first use consists of the calculation of the root around a point for the defined function.

```
> fzero(Myfunction,x0)
```

In this use, the given function is `Myfunction`, and the point is defined by `x0`.

In the second use, the function finds the root within a given interval.

```
>fzero(Myfunction, [x0, x1])
```

In this use, the given function is `Myfunction` and the root is searched between the points `x0` and `x1`. The function can be defined using an anonymous function as in the following example.

Example 4-3. Write code to find the root of 2 *cos* (*x*). The code should separately find the root around $\frac{\pi}{4}$ and the root within the interval of π and 2π. The code should print the values in terms of degrees on the screen.

Solution 4-3. The following code can be used to find the roots.

Example4p3.m

```
%Example4p3
%This code uses fzero function
Myfunc = @(x) 2*cos(x); % function
x0 = pi/4; % around this point
FirstKind = fzero(Myfunc,x0); % the usage
CorrespondingDegree1 = (FirstKind)*(180/pi);
fprintf("1st root is %d degree\n",CorrespondingDegree1)
interval = [pi, 2*pi];% interval for 2nd usage
SecondKind = fzero(Myfunc,interval);
CorrespondingDegree2 = (SecondKind)*(180/pi);
fprintf("2nd root is %d degree\n",CorrespondingDegree2)
```

As shown in the code, the values of the corresponding roots are converted from radians to degrees by simply multiplying these values by $180/\pi$.

Once the code is executed, the following output will be printed on the screen.

```
> Example4p3
1st root is 90 degree
2nd root is 270 degree
>
```

Bisection Method

The bisection method can be implemented by creating a user-defined function. There is no built-in function like fzero in MATLAB for the bisection method. The algorithm in the following example can be used for the implementation of this method. In the bisection method, the boundary points should be picked carefully. Due to the nature of the method, the defined function must have different signs at the boundaries (at a and b as presented in the code).

Example 4-4. Write a code to find the root of $f(x) = x^3 - 3x + 1$ between 0 and 1. The code should calculate the value up to a tolerance given by 10^{-10} using the bisection method.

Solution 4-4. The following code can be used to find this root.

Example4p4.m

```
%Example4p4
%This code uses bisection algorithm
format long
tolerance=1e-10;
a = 0.0; b = 1.0;
f=@(x) x^3-3*x+1;% function is defined here
Result = MyBisect (f,a,b,tolerance);
fprintf('Root is = f ( %d ) = %d \n',Result,f(Result));
function [c] = MyBisect (f,a,b,tolerance)
c = (a + b)/2;
while (abs(f(c)) > tolerance)
    c = (a + b)/2;
    if ( f(a)*f(c) < 0 )
       b = c;
    else
       a = c;
    end
end
end
```

Once the code is executed, the following output will be printed on the screen.

```
> Example4p4
Root is = f ( 3.472964e-01 ) = -4.365619e-11
>
```

Newton's Method

In some textbooks, Newton's method is referred to as the Newton–Raphson method. There is no built-in function in MATLAB to calculate a root by using this method, but its algorithm is quite easy to implement.

$$x_{n+1} = x_n - \frac{f(x_n)}{f'(x_n)}$$

The function must be defined by the user to calculate the root using this algorithm.

Example 4-5. Write code to find the root of $f(x) = x^3 - 3x + 1$ with an initial value of 0.5. The code should calculate the value up to a tolerance given by 10^{-10} by using Newton's method.

Solution 4-5. The following code can be used to illustrate Newton's method.

Example4p5.m

```
%Example4p5
%This code uses Newton's algorithm
format long
tolerans=1e-10;
x0 = 0.5;
F = @(x) x^3-3*x+1; %Function written here
Derivat = @(x) 3*x^2-3; % Its derivative should be here
%Make sure Derivat(x0) is NOT equal to zero
x0 = MyNewton (F,Derivat ,x0 ,tolerans);
fprintf('Root = f ( %d ) = %d \n',x0,F(x0));
```

```
function x0 = MyNewton (F,Derivat ,x0 ,tolerans)
while (abs(F(x0)) > tolerans)
    Result = x0 - (F(x0) / Derivat (x0));
    x0=Result;
end
end
```

Once we run the code, the following output will be displayed on the screen.

```
> Example4p5
Root = f ( 3.472964e-01 ) = -2.220446e-16
>
```

Note that the initial value plays an important role in finding the root more quickly.

Secant Method

The secant method is yet another method for finding the root of an equation. The algorithm for this method is

$$x_{n+1} = x_n - \frac{(x_n - x_{n-1})}{(f(x_n) - f(x_{n-1}))} f(x_n)$$

The function must be defined by the user to calculate the root using this algorithm. The following example can be used to illustrate the secant method.

Example 4-6. Write a code to find the root of $f(x) = x^3 - 3x + 1$ with the first two initial values of 0 and 1. The code should calculate the value up to a tolerance given by 10^{-10} by using the secant method.

Solution 4-6. The following code can be used to find the root via the secant method.

Example4p6.m

```
%Example4p6
%This code uses the secant method
format long
tolerance=1e-10;
x0 = 0.0;
x1 = 1.0;
```

```
f = @(x) x^3-3*x+1; % function comes here
x1 = MySecant (f,x0,x1 ,tolerance);
fprintf('Root = f ( %d ) = %d \n',x1,f(x1));

function x1 = MySecant (f,x0,x1 ,tolerance)
while (abs(f(x1)) > tolerance)
    Result = x1 - ( (x1 - x0) / (f(x1)-f(x0)) ) * f(x1);
    x0 = x1;
    x1 = Result;
end
end
```

Once the code is executed, the following output will be shown on the screen.

```
> Example4p6
Root = f ( 3.472964e-01 ) = -4.440892e-16
>
```

Fixed-Point Iteration

The fixed-point iteration method is generally used for finding the roots of equations in the form $f(x) = 0$. From the given equation, the iteration's function must be derived. In this method, the given equation is rewritten in the form $x = g(x)$, where $g(x)$ is the iteration function. In the algorithm, an initial point x_0 is selected. Then, the iteration function is recursively processed by $x_{n+1} = g(x_n)$. If the function f is continuous and x_n converges to a number, say M, then M is a fixed point of the iteration function.

For a certain function, more than one equation can be picked as the iteration function. Converging to a number plays an important role in picking the iteration function. To grasp the idea of how to select the iteration function, take a look at the following example.

Example 4-7. Considering the function given by $f(x) = x^2 - x - 4$, find possible iteration functions $g(x)$ to find the roots of $f(x)$.

Solution 4-7. The iteration function $g(x)$ can be selected as follows.

a) $g(x) = x^2 - 4$

b) $g(x)=1+\frac{4}{x}$

c) $g(x)=\sqrt{x+4}$

Example 4-8. Write code to find the root of $f(x) = x^3 - 3x + 1$ with an initial value of 0.5. The code should calculate the value up to a tolerance given by 10^{-10} using the fixed-point iteration method.

Solution 4-8. The following code can be used to find the root using this method.

Example4p8.m

```
%Example4p8
%This code uses fixed-point iteration method
format long
tolerans=1e-10;
x0 = 0.5;
f = @(x) x^3-3*x+1;% function comes here
while (abs(f(x0)) > tolerans)
    Result =  (x0 ^ 3 + 1) / 3; %fixed point function
    x0 = Result;
end
fprintf('Root = f ( %d ) = %d \n',x0,f(x0));
```

Once the code is executed, the following output will be displayed on the screen.

```
> Example4p8
Root = f ( 3.472964e-01 ) = -5.213918e-11
>
```

If we compare the results obtained by the different methods given, Newton's method yields a slightly better result than the others.

Numerical Integration

MATLAB provides various built-in functions, shown in Table 4-2, to calculate integrals numerically.

Table 4-2. *Functions Used in Numerical Integration*

Function	Explanation
integral	Performs integration by using the global adaptive quadrature
integral2	Evaluates double integrals numerically
integral3	Evaluates triple integrals numerically
cumtrapz	Calculates cumulative integrals using the trapezoidal method
trapz	Calculates integrals using the trapezoidal method

Example 4-9. Write code to numerically calculate the integral of $2xe^{x^2}$ from 0 to 1 by using the function `integral`. Then, compare the numerical solution with the analytical solution. The code should display the numerical result, analytical result, and error on the screen.

Solution 4-9. The following code can be used to accomplish the given task.

Example4p9.m

```
%Example4p9
%This code uses integral function
format long
Myfun = @(x) 2*exp(x.^2).*x;
NumericSol = integral(Myfun,0,1);
AnalyticSol = exp(1)-1;
Error = abs(NumericSol - AnalyticSol);
fprintf('Numerical solution :%e\n',NumericSol)
fprintf('Analytical solution:%e\n',AnalyticSol)
fprintf('Error is           :%e\n',Error)
```

Once the code is executed, the following output will be printed on the screen.

```
> Example4p9
Numerical solution :1.718282e+00
Analytical solution:1.718282e+00
Error is           :4.440892e-16
>
```

Numerical Differentiation

In this section, numerical methods for solving ordinary differential equations are discussed.

Ordinary Differential Equations

Numerical solution of ordinary differential equations has important applications in a wide variety of areas in engineering and applied sciences. MATLAB provides numerous built-in functions that solve ordinary differential equations with an initial condition.

Ordinary differential equations (ODEs) with initial values can also be solved using Euler's method and the Runge–Kutta method. This section covers how to solve initial value problems of ODEs using the ODE solvers, Euler's method, and the fourth-order Runge–Kutta method.

ODE Solver Functions

MATLAB has numerous functions for solving ODEs. Although `ode45` is the first one to come to mind, the user should consider the type of the problem (stiff or nonstiff) when selecting the solver.

Table 4-3. *Common ODE Solvers in MATLAB*

Solver	Accuracy	Explanation
ode45	Medium	Most of the time, this solver should be the first one tried; used for nonstiff problems
ode113	Low to high	Used for nonstiff problems, and when ODE is expensive to evaluate
ode15s	Low to medium	Recommended when ode45 fails, and when solving differential algebraic equations for stiff problems
ode15i	Low	Solves fully implicit differential equations for stiff problems

Example 4-10. Write code to solve the following initial value problem given by

$$\frac{dy}{dx}=y-x,y(0)=\frac{2}{3},0\leq x\leq 5$$

using the ode45 solver, where the exact solution is $y=(x+1)-\frac{1}{3}e^x$. The code should print the solution at x = 5 (y(5)) to the screen.

Solution 4-10. The following code can be used to accomplish the given tasks.

Example4p10.m

```
%Example4p10
%This code uses ode45 function
format long
y = 2.0/3.0; %initial condition
x0=0; xF =5;
f = @MyFunc;
xval = [0,5];% 0<=x<=5
[t,Result]=ode45(f,xval,y);
Lengt = length(Result);
fprintf("ode45 Solution of y(5): %.10f\n", Result(Lengt));
y_exact = 5.0 + 1.0 - (1.0/3.0)*exp(5.0);
fprintf("Exact Solution of y(5): %.10f\n",y_exact);

function Res = MyFunc(x,y)
Res = y-x; %defined function
end
```

In the code, the precision is set to show 10 decimal digits and the function `ode45` is called by an anonymous function. Once we run the code, the following output is displayed on the screen.

```
> Example4p10
ode45 Solution of y(5): -43.4717851406
Exact Solution of y(5): -43.4710530342
>
```

Euler's Method

Euler's method is yet another way to find approximate solutions to initial value problems of ODEs of the following type:

$$\{\frac{dy}{dx}=f(x,y)\,y(a)=y_a$$

$$y_{i+1}=y_i+hf(x_i,y_i)$$

where h is the mesh size (in a uniform mesh) over the interval $[a, b]$.

Example 4-11. Write code to solve the initial value problem given by

$$\frac{dy}{dx}=y-x, y(0)=\frac{2}{3}, 0\leq x\leq 5$$

using Euler's method, where the exact solution is $y=(x+1)-\frac{1}{3}e^x$. The code should print the exact solution and the numerical solution of $y(5)$ on the screen. The mesh size (h) should be set to $h = 0.0005$.

Solution 4-11. The following code can be used to accomplish the given tasks.

Example4p11.m

```
%Example4p11
%This code uses Euler's method
format long
f = @(x,y) y-x; %defined function
y = 2.0/3.0; %initial condition
xval = linspace(0,5,10001);% 0<=x<=5
h = xval(2)-xval(1); %h is mesh size and it is uniform
```

```
for i=1:length(xval)
    y = y + f(xval(i),y)*h; %Euler formula
end
fprintf("Numerical Solution of y(5): %.10f\n", y)
y_exact = 5.0 + 1.0 - (1.0/3.0)*exp(5.0);
fprintf("Exact Solution of y(5)     : %.10f\n",y_exact)
```

In the code, the precision is set to show 10 decimal digits. Once the code is executed, the following output will be shown on the screen.

```
> Example4p11
Numerical Solution of y(5): -43.4334780673
Exact Solution of y(5)     : -43.4710530342
>
```

Runge–Kutta Method of Fourth Order

Runge–Kutta methods are another group of methods for solving initial value problems of ODEs. In this section, we consider the fourth-order (**RK4**) formula, which has the following form:

$$\{\frac{dy}{dx} = f(x,y)\, y(a) = y_a, a \le x \le b$$

$$\{K_1 = hf(x_i, y_i)\, K_2 = hf\left(x_i + \frac{1}{2}h, y_i + \frac{1}{2}K_1 h\right) K_3 = hf\left(x_i + \frac{1}{2}h, y_i + \frac{1}{2}K_2 h\right) K_4$$
$$= hf(x_i + h, y_i + K_3 h)\, y_{i+1} = y_i + \frac{1}{6}(K_1 + 2K_2 + 2K_3 + K_4)$$

Example 4-12. Write code to solve the initial value problem given by

$$\frac{dy}{dx} = y - x, y(0) = \frac{2}{3}, 0 \le x \le 5$$

using **RK4,** where the exact solution is $y = (x+1) - \frac{1}{3}e^x$. The code should print the exact solution and the numerical solution of $y(5)$ on the screen. The mesh size (h) should be set to $h = 0.0005$.

Solution 4-12. We can write the following code for the given tasks.

Example14p12.m

```
%Example 4p12
%This example solves a first-order ODE
%by using Runge-Kutta Method of fourth order
format long
f = @(x,y) y-x; %defined function
y = 2.0/3.0; %initial condition
xval = linspace(0,5,10001);% 0<=x<=5
h = xval(2)-xval(1); %h is mesh size and it is uniform
for i=1:length(xval)
    K1 = h*f(xval(i),y);
    K2 = h*f(xval(i) + 0.5*h, y + 0.5*K1*h);
    K3 = h*f(xval(i) + 0.5*h, y + 0.5*K2*h);
    K4 = h*f(xval(i) + h, y + K3*h);
    y = y + (1.0/6.0)*(K1 + 2.0*K2 + 2.0*K3 + K4);%RK4
end
fprintf("Numerical Solution of y(5): %.10f\n", y)
y_exact = 5.0 + 1.0 - (1.0/3.0)*exp(5.0);
fprintf("Exact Solution of y(5)    : %.10f\n",y_exact)
```

In this code, the precision is set to show 10 decimal digits. Once the code is executed, the following output will be displayed on the screen.

```
> Example4p12
Numerical Solution of y(5): -43.4703160436
Exact Solution of y(5)    : -43.4710530342
>
```

Problems

4.1. Given the values $x = [1, 4, 8, 9]$ and $y = [15, 30, 50, 60]$, calculate the value of 6 using the default method of the `interp1` function.

4.2. Write code to find the root of 3 *sin* (x). The code should find the root within the interval $\frac{\pi}{2}$ and $\frac{3\pi}{2}$. The obtained value should be displayed on the screen in degrees.

4.3. Write code to find the root of $f(x) = x^4 + 4x + 2$ between -1 and 1. The code should calculate the value within the tolerance given by 10^{-6} using the bisection method.

4.4. Write code to find the root of $f(x) = x^4 + 4x + 2$ with an initial value of -1. The code should calculate the value within the tolerance given by 10^{-6} using Newton's method.

4.5. Write code to find the root of $f(x) = x^4 + 4x + 2$ with an initial value of -0.5. The code should calculate the value within the tolerance given by 10^{-6} using the fixed-point iteration method.

4.6. Write code to calculate the integral of $sin(2x)$ from 0 to π numerically using the `integral` function. Then, compare the numerical solution with the analytical solution. The code should print the numerical result, analytical result, and the discrepancy on the screen.

4.7. Write code to solve the initial value problem given by

$$\frac{dy}{dx} = 2x^2 + 7x + 6 - y, y(0) = 3, 0 \le x \le 5$$

using Euler's method. The code should print the numerical solution of $y(5)$ on the screen. The mesh size (h) should be set to $h = 0.0005$.

CHAPTER 5

Applications in Simulation

In this chapter, we first cover how to generate random numbers. Then, flipping a coin, rolling a pair of dice, random walking, and traffic flow topics are covered. In each section, I provide illustrations of the relevant programming ideas and explain them when necessary.

Random Number Generation

To create sequences of pseudorandom numbers in MATLAB, one of the rand, randn, or randi commands can be used (Table 5-1). We can use the rng function to control the repeatability of the numbers.

***Table 5-1.** Random Generating Functions*

Function	Explanation	Example
rand(n) rand(n,m)	Creates uniformly distributed pseudorandom numbers between 0 and 1	> rand(2) ans = 0.8147 0.1270 **0.9058 0.9134**
randn(n) randn(n,m)	Creates normally distributed pseudrandom numbers	> randn(2) ans = 0.3188 -0.4336 -1.3077 0.3426

(*continued*)

I. Turk, *Practical MATLAB*, https://doi.org/10.1007/978-1-4842-5281-9_5

Table 5-1. *(continued)*

Function	Explanation	Example

randi(imax)
randi(imax,x,y)
	Creates uniformly distributed pseudorandom integers up to imax	`> randi(5,2,3)`
		`ans =`
		`5 5 3`
		`1 5 5`

Rng	Controls random number generation	`> rng('shuffle')`

randperm(n)
randperm(n,k)
	Creates row vector containing *k* unique permutation of integers from 1 to *n*	`> randi(5,2,3)`
		`ans =`
		`2 4 3 1`

In a session, we can get different values each time we run one of the `rand`, `randn`, or `randi` functions. If the session is closed and a new session is opened, though, the same values are repeated for the same function used. This happens due to the fact that the `rng` function uses the same default seed. The default seed is Mersenne Twister with seed 0. If the seed type is changed to shuffle, then we can get different values each time `rng` is called. However, the `randperm(n,k)` function creates *k* random permutation of integers from 1 to *n*.

Example 5-1. Create seven numbers between 2 and 8 using the `rand` command.

Solution 5-1. To get seven numbers, we can use a row vector. If we use the `rand` command, we get numbers between 0 and 1. By multiplying the numbers by 10 and then rounding them, the obtained numbers will be between 0 and 10. If we use mode 7, then all numbers will be between 0 and 6. After shifting the numbers by adding 2, we can get the required numbers. To show all the steps, the following code can be used. However, the complete task can be achieved by typing just a single line as written at the bottom of the code.

Example5p1.m

```
%Example5p1
%This code produces 7 numbers
MyNumbers = rand(1,7);
MyBigNumb = MyNumbers*10;
MyIntegers = round(MyBigNumb);
```

```
MyMod = mod(MyIntegers,7);
MyFinal = MyMod + 2;
disp(['The numbers are: ',int2str(MyFinal)])
%MyFinal=2+mod(round(10*rand(1,7)),7)
```

Once the code is executed, the following output will be displayed on the screen.

```
> Example5p1
The numbers are: 3  3  4  7  6  8  2
>
```

Example 5-2. Using the `randperm` function, create five integers between 1 and 20. Then, create another sequence of 10 numbers using the randomly selected numbers with the `randperm` function. The first two elements of the sequence should be the first number created by `randperm`. The third and fourth elements of the sequence should be the second number created by `randperm`. The same pattern should be applied to the rest of the elements of the sequence.

Solution 5-2. The following code can be used to accomplish the given task.

Example5p2.m

```
%Example5p2
%This code uses randperm function
Numbers = randperm(20,5);
Seq= Numbers([1,1,2,2,3,3,4,4,5,5]);
disp('Randomly Selected Numbers:');
disp(Numbers);
disp('Numbers of Sequence:');
disp(Seq);
```

Once the code is executed, the following output will be shown on the screen.

```
> Example5p2
Randomly Selected Numbers:
     3     9    19     5     1
Numbers of Sequence:
     3     3     9     9    19    19     5     5     1     1
>
```

Flipping a Coin

An experiment involving flipping a coin can be performed using the technique in the preceding section, or simply by using the `randi(2,1)` command as illustrated in Example 5-3.

Example 5-3. Write code that asks the user the number of times to flip the coin. Then the code should call a function to determine the number of tails and heads in a sequence. The code should print the number of tails and heads and the flipped sequence on the screen.

Solution 5-3. The following code can be used to accomplish the given task.

Example5p3.m

```
%Example5p3
%This code spins coin
Times=input('Number of times to flip the coin?\n');
[H,T,Sequence]=FlipCoin(Times)
function [H,T,Sequence]=FlipCoin(Times)
rng('shuffle')
n=1; H=0; T=0;
Sequence = zeros(1,n);
while n <= Times
      Result = randi(2,1);
      if Result == 1
          H=H+1;
          Cond = 'H';
      else
          T=T+1;
          Cond = 'T';
      end
      Sequence(n) = Cond;
      n=n+1;
end
Sequence = char(Sequence);
end
```

The code calls the `FlipCoin` function to get random integer numbers with the `randi(2,1)` command. This command tells MATLAB that one integer will be selected: 1 or 2. Therefore, each time a number is selected. If the number is 1, then this is regarded as heads (`H`) otherwise, the result is saved as `T`, which means tails. All of these results are saved to the `Sequence` variable as numbers corresponding to H and T from the ASCII table. Then once the counting is done, the `Sequence` variable is converted to `H` and `T` using the `char` function.

Once the code is executed, the following output will be shown on the screen.

```
> Example5p3
Number of times to flip the coin?
10
H =
     2
T =
     8
Sequence =
    'TTHTTHTTTT'
>
```

Rolling a Pair of Dice

Programming the rolling of a pair of dice or a single die can be done in several ways. One of the easiest ways is to use the `randi(6,1,2)` command (or `randi(6,1)` for a single die). In this way, two integers can be selected between 1 to 6. For an illustration, we can take a look at Example 5-4.

Example 5-4. Write code to simulate rolling a pair of dice using the `randi` command. The code should ask the user the number of times the pair of dice is rolled. Then all outcomes including the sum of each roll should be displayed on the screen.

Solution 5-4. To get two numbers to represent die, you can use the `randi(6,1,2)` command. To achieve the task, the following code can be used.

Example5p4.m

```
%Example5p4
%This code rolls die and prints the sum
N=input('Enter the number of times to roll\n');
[Results,Sums] = RollDie(N);
```

```
disp('The Rolled Die');
disp(Results);
disp('The Sums For Each Rolling');
disp(Sums);
function [Results,Sums] = RollDie(Number)
Results = zeros(Number,2);
Sums = zeros(Number,1);
for i=1:Number
    Die=randi(6,1,2);
    Sum=Die(1)+Die(2);
    Results(i,1:2)=Die;
    Sums(i)=Sum;
end
end
```

Once the code is run, the following output is obtained.

```
> Example5p4
Enter the number of times to roll
5
The Rolled Die
     6     1
     2     6
     5     1
     2     4
     4     5
The Sums For Each Rolling
     7
     8
     6
     6
     9
>
```

Example 5-5. Write code to simulate rolling a pair of dice. The code should ask the user whether or not the user wants to play. If the user wants to play, the code rolls a pair of dice. If the sum of the numbers is 7, the computer should output "You WON $100" and the obtained numbers and keep rolling. If the sum of the numbers is 10, the computer should output "You WON $50000" and the obtained numbers and keep rolling. For the rest of the sum possibilities, the program should stop running.

Solution 5-5. The following code can be used to accomplish the given task.

Example5p5.m

```
%Example5p5
%This code simulates a pair of dice
N=input('Enter 1 to roll a pair of dice\n');
while (N==1)
    Die=randi(6,1,2);
    switch sum(Die)
        case 7
            disp('You WON $100');
            fprintf('Numbers: %i and %i\n',...
                Die(1),Die(2));
        case 10
            disp('You WON $50000');
            fprintf('Numbers: %i and %i\n',...
                Die(1),Die(2));
        otherwise
            N=2;
            fprintf('Numbers: %i and %i\n',...
                Die(1),Die(2));
    end
end
```

Once the code is executed, the following output will be printed on the screen.

```
> Example5p5
Enter 1 to roll a pair of dice
1
You WON $100
Numbers: 2 and 5
Numbers: 2 and 1
>
```

As shown in the output, after running the program and entering 1, the program began to run. Then, the computer rolled 2 and 5, the sum of which is 7, and printed `You WON $100` and continued. On the second attempt, the rolled numbers are 2 and 1, the sum of which is 3. Because 3 is not one of the special cases identified, the code changed the value of variable `N`, and finished.

Random Walking

A random walk can be simulated by using the `rand` command as well. This section illustrates one-dimensional and two-dimensional cases separately.

Example 5-6. Write a code to simulate a one-dimensional walk where the directions picked randomly from either 1 or -1 to go either to the right or left for 20 steps.

Solution 5-6. Although the question is one-dimensional, to see the right and left steps clearly, the path will be drawn one step up toward the *y* direction in each step. The right step will be in red color and blue will indicate the steps to the left. The following code can be used to accomplish the given task.

Example5p6.m

```
%Example5p6
%The code simulates random walk in 1-d
close all; Y=1; P = 0;%position
Steps = 20; Step = 1;
for i=1:Steps
    W = rand;
    if W>0.5
        P = P + Step;
        St = linspace(P-1,P,21);
        plot(St,Y,'r*')
        hold on
    else
        St = linspace(P,P-1,21);
        P = P - Step;
        plot(St,Y,'b+')
        hold on
    end
```

```
    Y = Y + 1;
    pause(0.5)
end
grid on
xlabel('Red-->right | Blue-->left')
ylabel('Direction in each step')
title('Random Walk in Each Step')
disp(['Final Place :',num2str(P)])
```

Once the code is run, the output shown in Figure 5-1 is obtained.

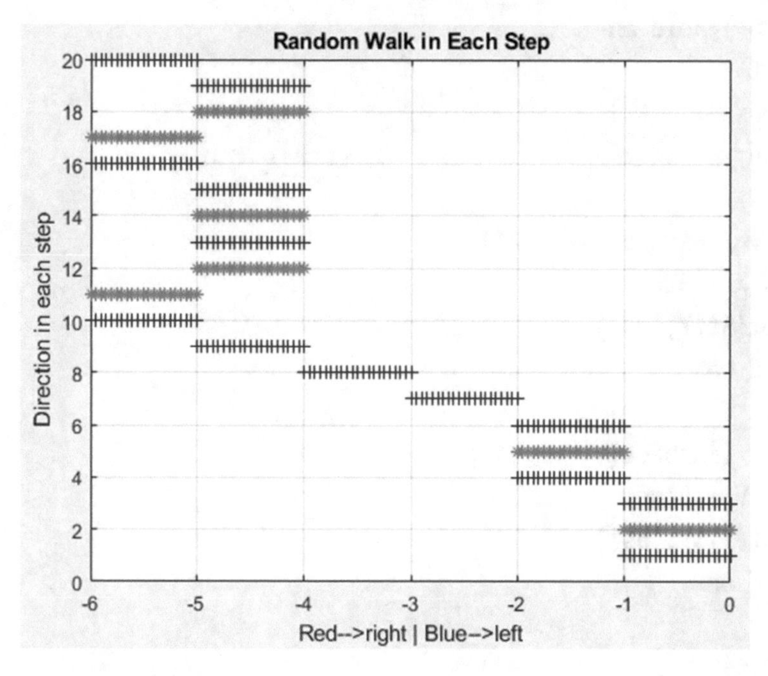

Figure 5-1. *Output of Example5p6*

The random walk starts by going left at the beginning. Then it goes right, then left again, and so on. At the end, the position is six steps left from the starting point. At the prompt, we also obtain the following result.

```
> Example5p6
Final Place :-6
>
```

Example 5-7. Write code to simulate a two-dimensional random walk. In each step, the computer can go just one step right in the x direction, or one step up in the y direction in 15 steps.

Solution 5-7. The following code can be used to accomplish the given tasks.

Example5p7.m

```
%Example5p7
%The code simulates random walk in 2-d
close all;clear all;P = 0;%position
Steps = 15;X=0;Y=0;Step = 1;
plot(X,Y,'co');hold on
format short
for i=1:Steps
    W = rand;
    if W<0.5
        St = linspace(X,X+1,21);
        X = X + Step;
        plot(St,Y,'r*')
        hold on
    else
        St = linspace(Y,Y+1,21);
        Y = Y + Step;
        plot(X,St,'bs')
        hold on
    end
    pause(0.5)
    disp([X,Y])
end
grid on
xlabel('Red-->Right | Blue-->Up')
ylabel('Direction in each step')
title('Random Walk in Each Step')
```

Once we run the code, the graphic shown in Figure 5-2 is obtained.

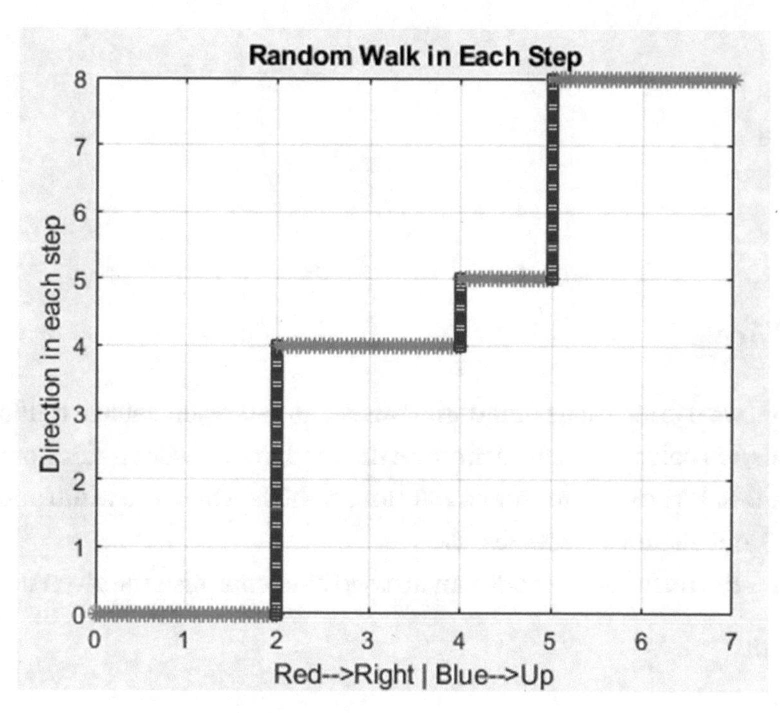

Figure 5-2. *Output of Example5p7*

As Figure 5-2 shows, the walk went right seven times and upward eight times. That total gives us the total number of steps in the code.

At the prompt, we also got the following after running the code.

```
> Example5p7
     1     0
     2     0
     2     1
     2     2
     2     3
     2     4
     3     4
     4     4
     4     5
```

```
     5     5
     5     6
     5     7
     5     8
     6     8
     7     8
>
```

Traffic Flow

In this section, we use an image function, `imagesc`, to represent vehicle traffic with color. In other words, we color different sections of the road to represent traffic flow.

Example 5-8. Write code to show a flow for a vehicle where the width and length of the road are 5 and 20 units, respectively.

Solution 5-8. The following code can be used to accomplish the given tasks.

Example5p8.m

```
%Example5p8
%This code shows traffic flow
rw=5;%road width
rl=25;%road length
road = zeros(rw,rl);
xpos=2;ypos=4;
road(ypos,xpos)=1;
imagesc(road)
axis equal
for t=0:20
    %clear previous vehicle
    road(ypos,xpos)=0;
    %update new position
    xpos=xpos +1;
    road(ypos,xpos)=1;
    %plot new position
    imagesc(road)
```

```
    Time=sprintf('Time: %i',t);
    title(Time)
    axis equal;
    pause(0.4);
end
```

As shown in the code, after moving to the right side in each iteration, the color of the vehicle road is assigned to 0. To move forward, the color of the road is assigned to 1. And each time, the colors are displayed with the `imagesc` function. This is done to display a traffic flow in a graphic form. Between each frame, the computer waits 0.4 seconds to see the simulation slowly.

Figure 5-3 shows the final frame of the simulation done with the code.

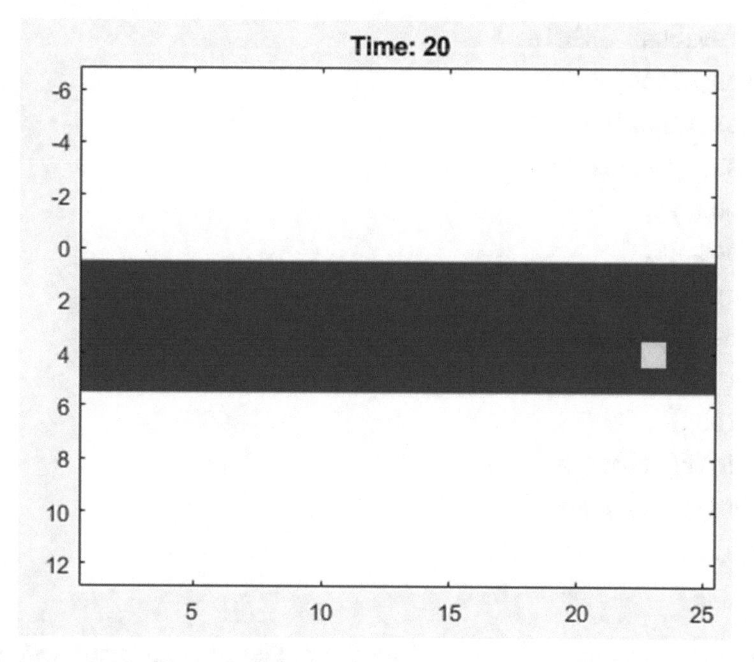

Figure 5-3. *Output of Example5p8*

Example 5-9. Write code to show a flow with two separate directions of traffic: going and coming. On each road, there should be one vehicle driving.

Solution 5-9. The following code can be used to accomplish the given task.

Example5p9.m

```
%Example5p9
%This code shows traffic flow
rw=11;rl=25;
road = zeros(rw,rl);
road(6,:)=1;
x1pos=2;y1pos=9;
x2pos=19;y2pos=3;
road(y1pos,x1pos)=1;
road(y2pos,x2pos)=1;
imagesc(road);axis equal
for t=0:15
    %clear previous vehicle
    road(y1pos,x1pos)=0;
    road(y2pos,x2pos)=0;
    %update new position
    x1pos=x1pos +1;
    x2pos=x2pos -1;
    road(y1pos,x1pos)=1;
    road(y2pos,x2pos)=1;
    %plot new position
    imagesc(road)
    Time=sprintf('Time: %i',t);
    title(Time);axis equal;
    pause(0.4);
end
```

Once the code is executed, the output shown in Figure 5-4 is displayed on the screen.

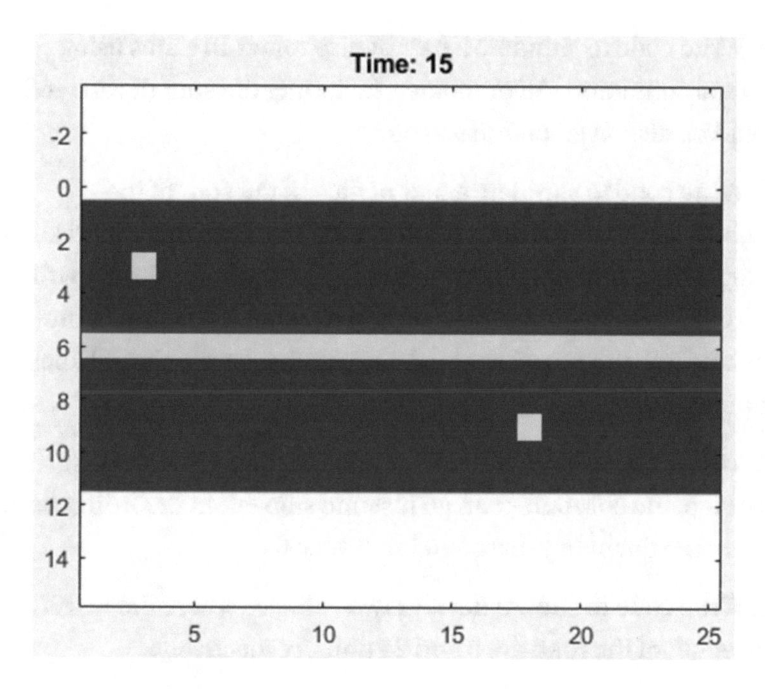

***Figure 5-4.** Output of Example5p9*

Problems

5.1. Create 10 numbers between 3 and 7 using the `randi` command.

5.2. Using the `randperm` function, create seven integers between 1 and 15. Then, create another sequence of 20 numbers using randomly selected numbers with the `randperm` function. The first three elements of the sequence should be the first number created by `randperm`. The fourth, fifth, and sixth elements of the sequence should be the second number created by `randperm`. The same pattern should be applied to the rest of the elements of the sequence.

5.3. Write code that flips a coin 20 times. The code should print the number of tails and heads obtained and the flipped sequence on the screen.

5.4. Write code to simulate a pair of dice rolled 10 times using the `rand` command. All outcomes, including the sum of each roll, should be displayed on the screen.

5.5. Write code to simulate a pair of dice. If the sum of the numbers is 5, the program should write "You GOT $200" and keep rolling. If the sum of the numbers is 12, the program should write "You GOT THE BIG PRIZE!!" and keep rolling. If the sum of the numbers is 8, the program should stop; otherwise it should keep running.

5.6. Write code to simulate two-dimensional random walking. In each step, the computer can go just one step left in the x direction, or one step down in y direction for 20 steps.

5.7. Write code to show a flow for two vehicles where the width and length of the road are 6 and 24 units, respectively.

CHAPTER 6

Basic Statistics and Data Analysis

This chapter begins with a brief discussion about basic statistical functions used in MATLAB. After that, sorting, searching, and processing data with Microsoft Excel files are presented.

Basic Statistics

MATLAB offers a great variety of features and ready-to-use, built-in functions in the area of statistics. In this book, I only introduce an outline of the topic with some basic statistical functions (Table 6-1).

Table 6-1. *Some of the Basic Functions Related to Statistics in MATLAB*

Function	Explanation	Example
max()	Returns the maximum element	`max([1,34,21,5])`
mean()	Returns the average or mean value	`mean([1,34,21,5])`
median()	Returns the median value	`median([1,34,21,5])`
min()	Returns the smallest value	`min([1,34,21,5])`
mode()	Returns the most frequently occurring value	`mode([1,34,5,21,5])`
std()	Returns the standard deviation	`std([1,34,5,21,5])`
var()	Returns the variance	`var([1,34,5,21,5])`

I. Turk, *Practical MATLAB*, https://doi.org/10.1007/978-1-4842-5281-9_6

Example 6-1. Create a 3 × 4 matrix randomly by using the function randi(). Then, calculate the maximum value, mean value, and standard deviation of each column of the matrix.

Solution 6-1. The following code can be used to create the matrix and find the desired values for each column of the matrix.

Example6p1.m

```
%Example6p1
%This code finds statistical values
A=randi(100,3,4);
%Maximum number is set as 100
Maximum = max(A);
MeanVal = mean(A);
StanDev = std(A);
disp(['The maximum numbers are ',...
    num2str(Maximum)])
disp(['The mean numbers are ',...
    num2str(MeanVal)])
disp(['The standard deviations are ',...
    num2str(StanDev)])
```

Once the code is executed, the following output will be displayed on the screen.

```
> Example6p1
The maximum numbers are 84  92  97  98
The mean numbers are 37.6667          57       60.3333        56.3333
The standard deviations are 40.2782        44.441       32.5167        36.8556
>
```

Data Analysis

This section presents various sorting and searching functions, along with data processing examples.

Sorting and Searching

Sorting in ascending or descending order can easily be achieved via the `sort` function.

Example 6-2. Consider the data given by 23, 70, 86, 5, 120, 50, -9, 61, and 100. Write a code to sort the given values both in ascending and descending order.

Solution 6-2. The data can be introduced as a vector, and then can be sorted as shown here at the prompt.

Example6p2.m

```
%Example6p2
%This code sorts data
A=[23,70,86,5,120,50,-9,61,100];
DescendSort = sort(A,'descend');
AscendSort = sort(A,'ascend');
disp('Descending Sort:');
disp(DescendSort);
disp('Ascending Sort:');
disp(AscendSort);
```

Once the code is run, the following output is obtained.

```
> Example6p2
Descending Sort:
   120   100    86    70    61    50    23     5    -9
Ascending Sort:
    -9     5    23    50    61    70    86   100   120
>
```

In some cases, multiple sets of data, as in a matrix, need to be sorted. This case is shown in Example 6-3.

Example 6-3. Randomly create a 3 × 5 matrix by using the `randi()` function. Then, sort each column and each row of the matrix in descending order.

Solution 6-3. The following code can be used to sort the columns and rows of the matrix.

Example6p3.m

```
%Example6p3
%This code sorts rows and columns
A=randi(100,3,5);
SortColumn=sort(A,1);
SortRow=sort(A,2);
disp('Original Matrix');
disp(A);
disp('Column Sorted');
disp(SortColumn)
disp('Row Sorted');
disp(SortRow)
```

Once the code is run, the following output is obtained.

```
> Example6p3
Original Matrix
    94    29    87   100    58
    46     1    40    54    65
    32    61    26    96    27
Column Sorted
    32     1    26    54    27
    46    29    40    96    58
    94    61    87   100    65
Row Sorted
    29    58    87    94   100
     1    40    46    54    65
    26    27    32    61    96
>
```

As shown in the output, if the command `sort` is used in the form `sort(Matrix,1)`, sorting is performed for columns; if `sort(Matrix,2)` is used, then sorting is performed for rows.

Searching is also a straightforward task with MATLAB. Typically, there are three different possibilities available for searching. A number can be searched in a vector or matrix. A group of letters or a string can be searched in a string cluster. Finally, we can search some information within a dataset. We will examine each of these cases with an example.

Example 6-4. Consider the matrix given by Mat = [32,98,17,71,67; 39,71,65,28,17; 77,49,71,68,71]. Write code that searches for 71 within this matrix.

Solution 6-4. The following code can be written to perform the search.

Example6p4.m

```
%Example6p4
%This code find indexes in a matrix
Matrx =[32,98,17,71,67;...
        39,71,65,28,17;...
        77,49,71,68,71];
Key = 71;
Indexing = find(Matrx==Key);
disp('Its indexes are: ')
disp(Indexing);
```

Once the code is run, the following output is obtained.

```
> Example6p4
Its indexes are:
     5
     9
    10
    15
>
```

As shown, MATLAB returns the index values where the number 71 occupies a place. The fifth, ninth, tenth, and fifteenth elements of the matrix have a value of 71. From this example, we see that the order of indexing goes column by column.

Example 6-5. Consider the string given by `12345 ABcde Antonio`. Write code that looks for the string `antonio` without considering case-sensitivity.

Solution 6-5. MATLAB is a case-sensitive language. Therefore, the strings `Antonio` and `antonio` are two different strings. The following code can be used to search for the desired string.

Example6p5.m

```
%Example6p5
%This code finds string
Text = '12345 ABcde Antonio';
Scr = 'antonio';
NewText = upper(Text);
NewScr= upper(Scr);
Findit = strfind(NewText,NewScr);
disp('Place of string: ');
disp(Findit);
```

The strings can be defined in MATLAB as vectors containing these strings, as shown in the preceding code. After running the code, the following output will be obtained.

```
> Example6p5
The string is
    13
>
```

This output states that the searched string is embedded in the bigger string, and it starts from the thirteenth element of the vector. If the string did not exist, then we would have gotten an empty set [ ], meaning that the searched word could not be found.

Example 6-6. In MATLAB, there is a dataset available for use with the name `hospital`. A portion of these data are shown in Figure 6-1.

```
mmand Window
>> load hospital
>> hospital
hospital =
              LastName          Sex       Age     Weight
    YPL-320   'SMITH'           Male      38      176
    GLI-532   'JOHNSON'         Male      43      163
    PNI-258   'WILLIAMS'        Female    38      131
    MIJ-579   'JONES'           Female    40      133
    XLK-030   'BROWN'           Female    49      119
    TFP-518   'DAVIS'           Female    46      142
    LPD-746   'MILLER'          Female    33      142
    ATA-945   'WILSON'          Male      40      180
    VNL-702   'MOORE'           Male      28      183
```

Figure 6-1. *A screenshot from the hospital dataset*

Let us load these data into the workspace. Then, search the dataset to find the last name DIAZ, and other relevant information. After that, find the people who are 50 years old.

Solution 6-6. To find DIAZ, we need to search the column LastName, which contains strings. To search for the number 50, we need to work with the column Age. The following code can be used to perform the given tasks.

Example6p6.m

```
%Example6p6
%This code finds data
load hospital
IsThere=ismember(hospital.LastName,'DIAZ');
index = find(IsThere);
fprintf('The person last name DIAZ is\n')
hospital(index,:)
fprintf('Information having age 50:\n')
hospital(find(hospital.Age==50),:)
```

In this code, the command `ismember` determines whether DIAZ is a member of the cell `hospital.LastName` or not. It returns 1 if it is the case, and 0 if it is not. By using the `find` command, it is possible to find out the index of the required information.

Once the code is executed, we will obtain the following output.

```
> Example6p6
The person last name DIAZ is
ans =
           LastName    Sex     Age   Weight    Smoker    BloodPressure
    BEZ-311 'DIAZ'     Male    45    172       true      136         93

              Trials
    BEZ-311   [1×0 double]
Information having age 50:
ans =
              LastName          Sex     Age    Weight    Smoker
    XBA-581   'ROBINSON'        Male    50     172       false
    DAU-529   'REED'            Male    50     186       true

              BloodPressure      Trials
    XBA-581   125         76     [1×3 double]
    DAU-529   129         89     [         22]
>
```

Data Processing

This section presents how to pull out and process information from a Microsoft Excel data file via an example.

Example 6-7. Write code to get the information from the Excel file `DataProcessing1.xlsx` (Figure 6-2).

LastName	Weight(kg)	Lifting Capacity(kg)	Eating Capacity(kg)
DIAZ	102	120	10
MARCELLO	99	110	15
KORKMAZ	92	100	20
BILL	80	140	22
ANTONIO	85	135	17
HATTON	115	125	14
WATSON	110	120	16
KELLY	85	110	18
DUNCAN	100	140	20
WESTBROOK	95	145	12

Figure 6-2. *The content of the file* `DataProcessing1.xlsx`

Acquire the data from the second, third, and fourth columns, including their corresponding titles. Plot these data in a bar graph.

Solution 6-7. The following code can be used to accomplish the assigned task.

Example6p7.m

```
%Example6p7
%This code plots graphics from an Excel file
DataFile = importdata('DataProces.xlsx');
NewVar1=(DataFile.textdata.Sheet1{1,2});
NewVar2=(DataFile.textdata.Sheet1{1,3});
NewVar3=(DataFile.textdata.Sheet1{1,4});
bar(DataFile.data.Sheet1)
grid on
title('Sample Data')
xlabel('Number of Persons');
ylabel('kg');
legend(NewVar1,NewVar2,NewVar3);
```

Here, we can get the column titles from the variables `NewVar1`, `NewVar2`, and `NewVar3`. Once the code is executed, we will see the graphic result shown in Figure 6-3.

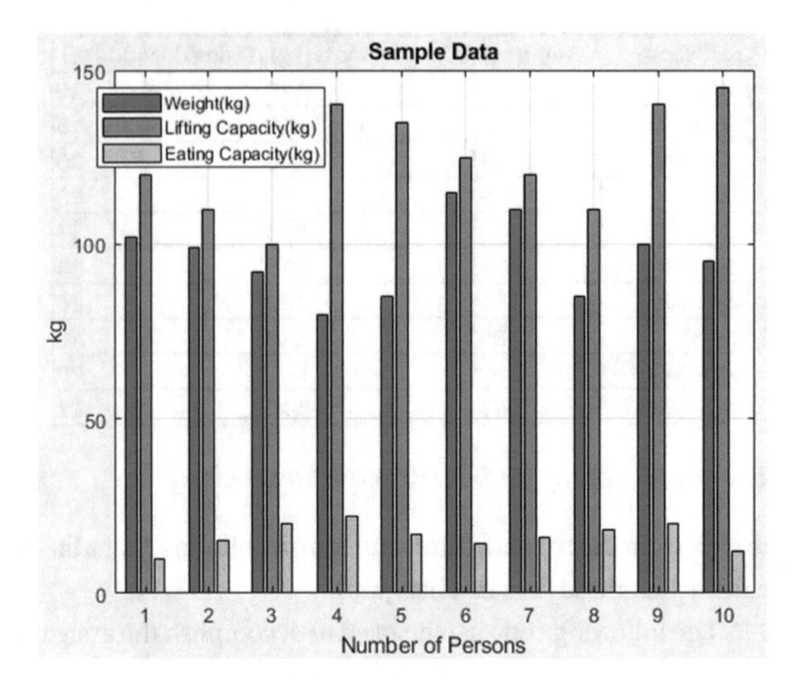

Figure 6-3. *The output obtained by Example6p7*

In MATLAB, you can work with comma-separated value (`.csv`) files as well.

Example 6-8. Write code to print the first five rows of the data from the `outages.csv` file from MATLAB.

Solution 6-8. The following code can be used to accomplish the given task.

Example6p8.m

```
%Example6p8
%This code works with csv data
T = readtable('outages.csv');
Y=head(T,4); % show first 4 rows of table
disp(Y)
```

Once the code is run, the following output is obtained.

```
> Example6p8
Region       OutageTime         Loss     Customers    RestorationTime     Cause
________     ________________   ______   __________   ________________    _________

'SouthWest'  2002-02-01 12:18   458.98   1.8202e+06   2002-02-07 16:50   'winter storm'
'SouthEast'  2003-01-23 00:49   530.14   2.1204e+05    NaT               'winter storm'
'SouthEast'  2003-02-07 21:15   289.4    1.4294e+05   2003-02-17 08:14   'winter storm'
  'West'     2004-04-06 05:44   434.81   3.4037e+05   2004-04-06 06:10   'equipment fault'
>
```

Problems

6.1. Create a 4 × 5 matrix randomly by using the `rand()` function. Then, calculate the maximum value, mean value, and standard deviation for each column of the matrix.

6.2. Consider the data 30, 45, 100, 65, 98, 45, 61, and 10. Write a code to sort the given data both in ascending and descending order.

6.3. Randomly create a 2 × 6 matrix by using the `randi()` function. Then, sort each column and each row of the matrix in descending order.

6.4. Consider the matrix given by Mat = [41, 45, 100, 65, 41, 45, 61, 10]. Write code that searches for `41` within this matrix.

6.5. Consider the string given by `5361 Sen Antonio Ben Banderas`. Write code that looks for the string `anderas` without considering case-sensitivity.

CHAPTER 7

Data Visualization and Animation

MATLAB provides very powerful techniques to visualize data. In this chapter, we look at how to visualize data and create animations. Visualizing data was already touched on in Chapter 1. Therefore, the animation techniques are emphasized in this chapter. Animations can be created by using three basic techniques in MATLAB. These methods can be summarized as updating coordinates, applying transformation to objects, and creating movies.

Data Visualization

We already learned how to visualize two-dimensional data Chapter 1 for both single plots and multiple plots in a figure. In this section we learn how to visualize three-dimensional data (Table 7-1).

I. Turk, *Practical MATLAB*, https://doi.org/10.1007/978-1-4842-5281-9_7

Table 7-1. *Some Functions Related to 3-D Plotting*

Function	Description
plot3(x,y,z)	Creates a 3-D line plot
bar3	Plots 3-D bar graphs
comet3(x,y,z)	Plots an animated 3-D graph
ezmesh	Visualizes the function in 3-D
ezplot3	A 3-D parametric curve plotter
mesh(x,y,z)	Creates a meshed-surface plot
pie3	Draws a 3-D pie chart
scatter3	Is a 3-D scatter plot function
stem3	Creates a 3-D stem plot
surf	Plots a 3-D shaded surface
waterfall	Creates a waterfall plot

Example 7-1. Consider the three-dimensional function given here.

$$0 \leq t \leq 4\pi, x = \sin(\mathrm{t}), y = \cos(\mathrm{t}), z = \mathrm{t}.$$

Plot the graph of *x, y,* and *z* in a single figure.

Solution 7-1. The following code can be used to accomplish the given task.

Example7p1.m

```
%Example7p1
%This code plots 3-D
close all;
t = linspace(0,8*pi,1000);
x=sin(3*t);y=cos(2*t)-5;z=3*t;
plot3(x,y,z)
xlabel('sin(t)');
ylabel('cos(t)');
zlabel('t');
grid on
title('3-D Plot')
```

In this code, the `close all` command closes all the active figures drawn when it is called. Once the code is run, the output shown in Figure 7-1 is displayed.

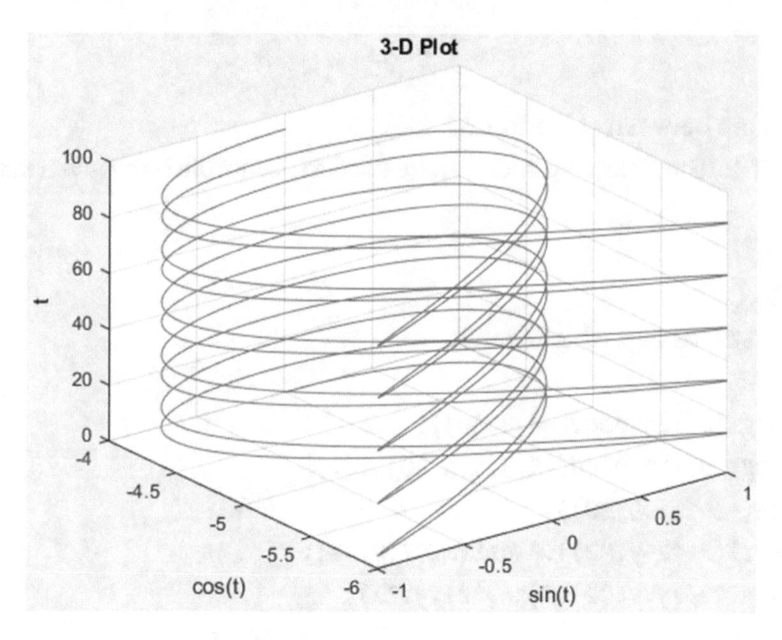

Figure 7-1. *Output of Example7p1*

There are two important functions that produce values with two variables in MATLAB, namely, `peaks` and `meshgrid`.

The function `peaks(x,y)` produces a 49 × 49 matrix by default via Gaussian distribution. The function `meshgrid(x,y)` replicates the grid vectors *x* and *y* to produce a full grid.

Using the functions `[X,Y,Z] = peaks(x,y,z)` and `[X,Y,Z] = meshgrid(x,y,z)`, it is possible to create three variables, as well.

Example 7-2. Write code that plots two different functions onto two separate figures. The first plot belongs to the following function (which plots the Mexican hat):

$$\frac{\sin\left(\sqrt{x^2+y^2}\right)}{\sqrt{x^2+y^2}}$$

where x and y are between -8 and 8. The second function (which plots a cowboy hat) is

$$\frac{\sin\left(3x^2+2y^2\right)}{x^2+y^2}$$

where x and y are between -1.5 and 1.5.

Solution 7-2. The following code can be used to accomplish the given task.

Example7p2.m

```
%Example7p2
%This code uses meshgrid and surf
close all;
[x,y] = meshgrid(linspace(-8,8,30));
[xx,yy]=meshgrid(linspace(-1.5,1.5,30),...
    linspace(-1.5,1.5,30));
z1 = sin (sqrt(x.^2+y.^2))./sqrt(x.^2+y.^2);
z2 = sin(3*xx.^2+2*yy.^2)./(xx.^2+yy.^2);
figure(1)
surf(z1), shading flat
title('Mexican Hat');
figure(2)
surf(z2), shading faceted
title('Cowboy Hat');
% other possible shadings: flat, faceted, interp
```

Once the code is run, the resulting output is shown in Figures 7-2 and 7-3.

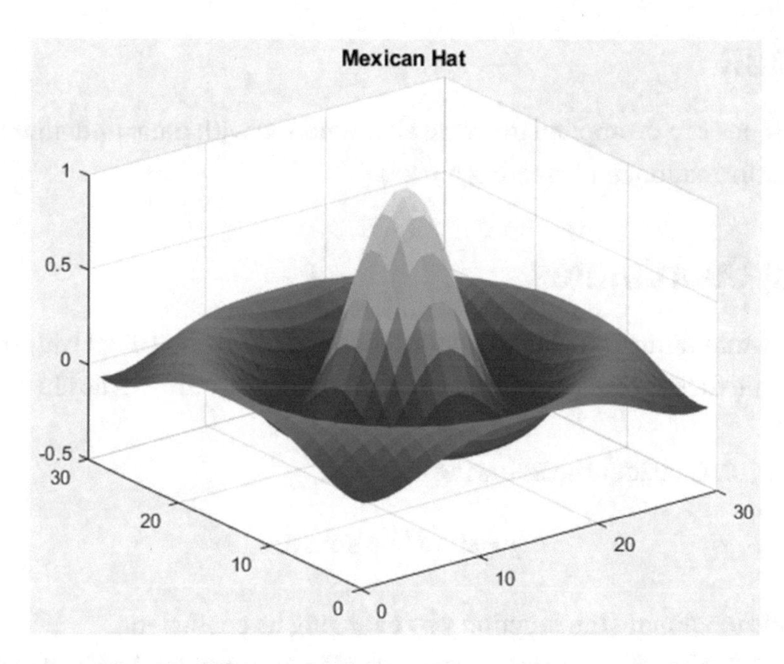

Figure 7-2. *First output of Example7p2*

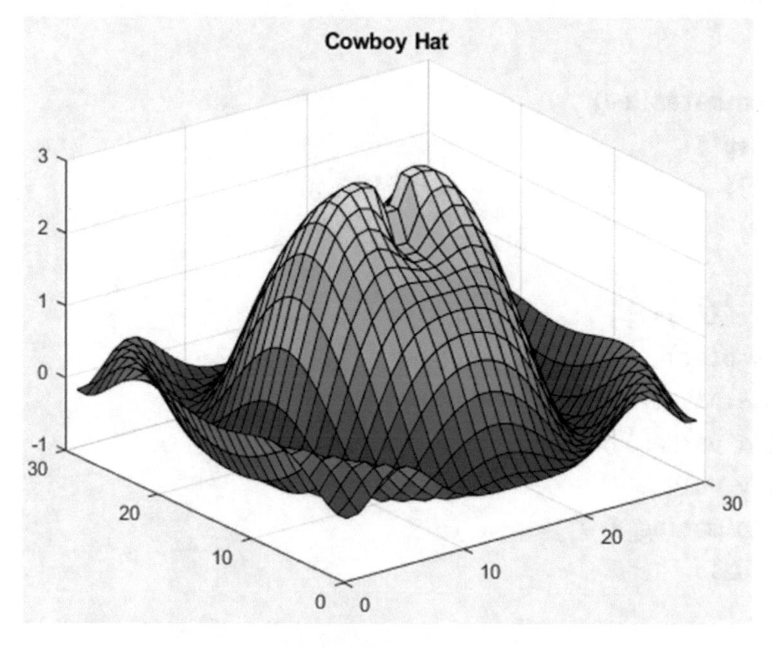

Figure 7-3. *Second output of Example7p2*

Animation

Three techniques can be applied to create an animation with data: updating coordinates, applying transformation, and creating movies.

Updating Coordinates

In this type of animation, the object properties are updated and called within a loop. Most of the time these properties are the data of x and y coordinates as illustrated in Example 7-3.

Example 7-3. Consider the following function.

$$y = \sin\left(x^2\right), 0 \le x \le 3\pi$$

Write code to animate the function y by changing its coefficient.

Solution 7-3. The following code can be used to accomplish the given task.

Example7p3.m

```
%Example7p3
%This code animates 2-D
x = 0:0.05:3*pi;
y = sin(x.^2);
N = length(x);
for i=1:N
    y_plot = (i/N)*y;
    plot(x,y_plot)
    axis([0,3*pi,-1,1]);
    xlabel('x values')
    ylabel('y values')
    title('Animating 2-D')
    pause(0.02)
end
grid on
```

In the preceding code, the function *y* is plotted with coefficients starting from $\frac{1}{189}$ to 1. After each plotting in the loop, the pause(0.02) command pauses the computer for 0.02 seconds to see the changes in each frame visually.

Once the code is run, the plot is animated. The plot shown in Figure 7-4 is the final frame of the animation.

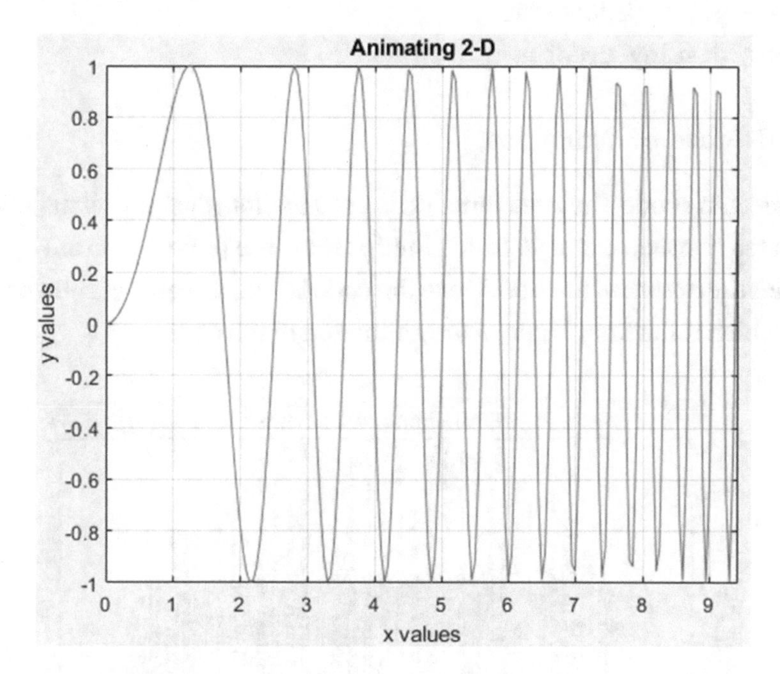

***Figure 7-4.** A snapshot of the output of Example7p3*

Example 7-4. Write code that uses the drawnow function to show the animation. The code should draw lines in a circle.

Solution 7-4. The following code can be used to accomplish the given task.

Example7p4.m

```
%Example7p4
%This code animates 2-D
N=100;
Angle = linspace(-pi,pi,N);
xc = sin(Angle);yc = cos(Angle);
plot(xc,yc);axis equal
xt = [1 1 1 1];yt = [0 0 0 0];
```

```
hold on
t = area(xt,yt); % initialize flat triangle
for j = 1:N
    xt(j) = xc(j); % determine new vertex value
    yt(j) = yc(j);
    t.XData = xt; % update data
    t.YData = yt; % update data
    drawnow % display updates
end
title('Final Frame of Animation')
```

In the preceding code, the `area` function is used to plot *yt* values versus *xt* values and fills the area between 0 and *yt*. Inside the `for` loop, the plotted data are updated and displayed with the `drawnow` function. Once the code is run, the output shown in Figure 7-5 is obtained, which is the final frame of the animation of the code.

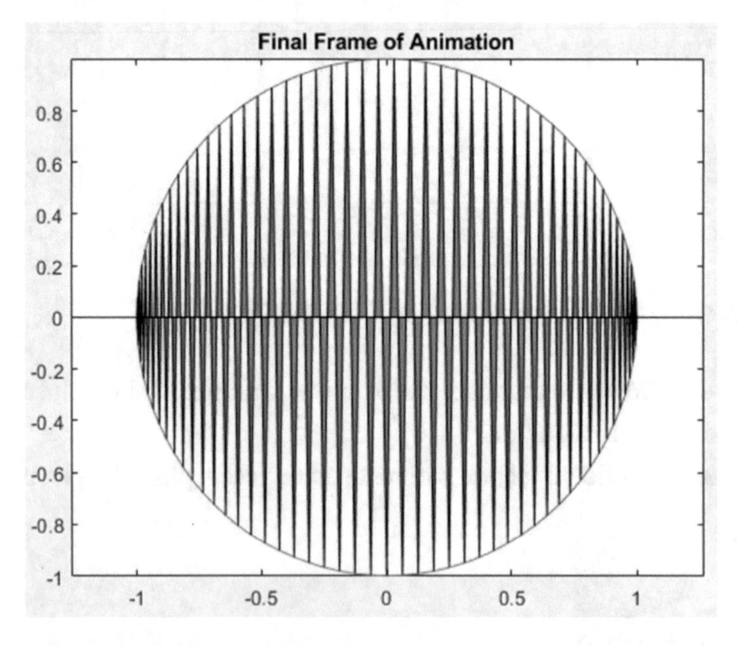

Figure 7-5. *A snapshot of the output of Example7p4*

Applying Transformation

In this technique, transformation is applied to objects. The function `hgtransform` is used to create the transform object.

Example 7-5. Write code to animate a 3-D star around the *z*-axis while the star is scaled from a large size to a smaller size by using the hgtransform.

Solution 7-5. The following code can be used to accomplish the given task.

Example7p5.m

```
%Example7p5
%This code uses hgtransform function
ax = axes('XLim',[-1.5,1.5],...
    'YLim',[-1.5,1.5],'ZLim',[-1.5,1.5]);
view(3) %sets for 3-d view
grid on
[x,y,z] = cylinder([.1 0]);
h(1) = surface(x,y,z,'FaceColor','yellow');
h(2) = surface(x,y,-z,'FaceColor','cyan');
h(3) = surface(z,x,y,'FaceColor','magenta');
h(4) = surface(-z,x,y,'FaceColor','green');
h(5) = surface(y,z,x,'FaceColor','blue');
h(6) = surface(y,-z,x,'FaceColor','red');
t = hgtransform('Parent',ax);
set(h,'Parent',t)
Rz = eye(4);Sxy = Rz;
for r = 1:.1:2*pi
    Rz = makehgtform('zrotate',r);
    % Scaling matrix
    Sxy = makehgtform('scale',r/5);
    set(t,'Matrix',Rz/Sxy)
    drawnow
    pause(0.02)
    if r == 3.6
        f = getframe;
    end
end
imshow(f.cdata)
title('Frame at r=3.6')
```

Once the code is run, the output shown in Figure 7-6 is obtained.

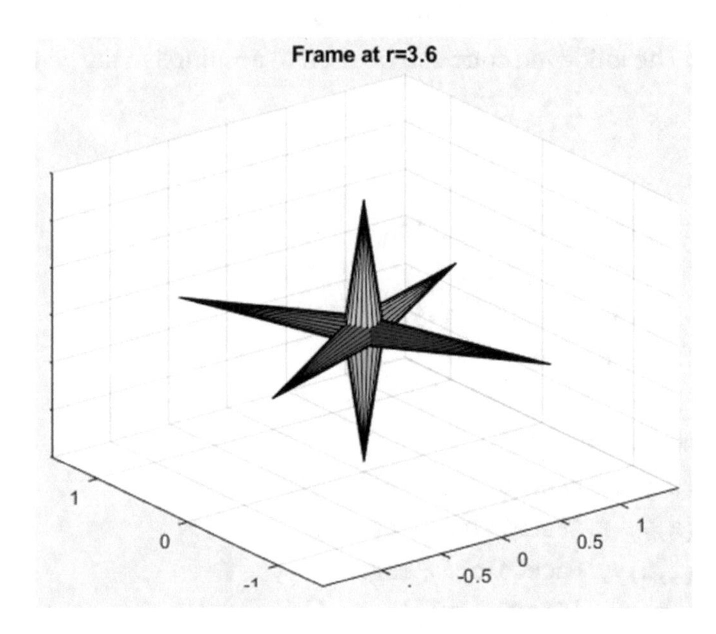

Figure 7-6. *Output of Example7p5*

Creating Movies

Creating a movie is another method to animate the data. In this method, the picture of each drawing in each iteration is obtained using the `getframe` function. Then, using the `movie` function, the animation is created.

Example 7-6. For the following function given by

$$y = \sin(t), 0 \le t \le 2\pi,$$

write code to animate the function ***y*** as a movie.

Solution 7-6. The following code can be used to accomplish the given task.

Example7p6.m

```
%Example7p6
%This code creates a movie
t=linspace(0,2*pi,1000);Count=1;
figure('Name',...%'NumberTitle','off')
    'Original Draw','Menu','none')
```

```
for freq=0:0.1:2*pi
        y=sin(freq*t);
        plot(t,y);
        xlabel('2 pi');ylabel('Results');
        axis([0,2*pi,-1,1])
        Script1=sprintf('y(t)=sin(%.1f t)',freq);
        title('Sinusoidal Function');
        text(1,0.5,Script1)
        M(Count)=getframe;Count=Count+1;
end
figure('Name',...
    'Playing Created Animation twice','Menu','none')
movie(M,2);
```

Once the code is run, the output is shown in Figure 7-7.

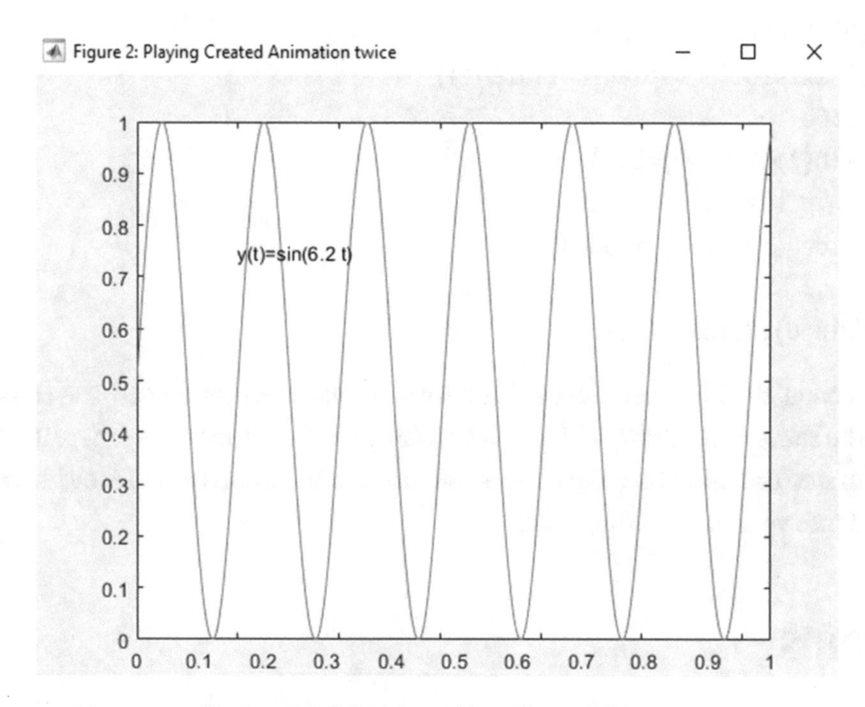

Figure 7-7. *The second output of the program Example7p6*

When you want to save the movie that you created, use the functions `writeVideo` and `VideoWriter`.

Example 7-7. Create a movie by using the functions peaks and surf. Then, save the movie as an *.avi file.

Solution 7-7. The following code can be used to create the *.avi file.

Example7p7.m

```
%Example7p7
%This code saves the movie as *.avi
myVideo = VideoWriter('myfile.avi');
uncompressedVideo = VideoWriter('myfile.avi',...
    'Uncompressed AVI');
myVideo.FrameRate = 40;
myVideo.Quality = 100;
open(myVideo); % Open file to write
L = peaks;
surf(L);
axis tight manual
set(gca,'nextplot','replacechildren');
for m = 1:60
   surf(sin(2*pi*m/20)*L,L)
   frame = getframe;
   writeVideo(myVideo, frame);
end
close(myVideo);%close file
```

In this code, the function VideoWriter creates a video writer object. The class of this object is the same as its name, VideoWriter. Using the function writeVideo, the frames are written into that file. Once the code is executed, a file named myfile.avi is created and saved into your current directory.

Problems

7.1. Consider the three-dimensional function given here.

$$0 \leq t \leq 2\pi, x = \sin(2t), y = \cos(3t), z = 5t$$

Plot the graph of *x*, *y*, and *z* in a single figure.

7.2. Consider the following function.

$$y = \sin\left(x^3\right), 0 \le x \le \pi$$

Write code to animate the function *y* by changing its coefficient.

7.3. Write code that uses the `drawnow` function to show the animation. The code should draw lines in a square.

7.4. Write a code to animate a 3-D star around the *y*-axis while the star is scaled from a small size to a larger size using the `hgtransform` function.

7.5. For the following function given by

$$y = \sin(2t), 0 \le t \le \pi,$$

write code to animate the function *y* twice as a movie and save the created animation.

CHAPTER 8

Computational Biology

Computational biology is a division of applied science that combines application of theoretical methods, mathematical modeling, and computer science to solve biological, ecological, social, and behavioral problems. In this chapter, we see some of its applications written with MATLAB code. Bacterial growth and population models are illustrated with a few examples. Then, topics including host–parasitoid models, bioinformatics, predator–prey models, and epidemic models are presented.

Bacterial Growth and Population Models

Bacterial growth or a simulation of a population can be modeled using differential equations. Ideally, bacterial growth can be represented as $\frac{dN}{dt} = \alpha N$, where N is the number of cells at the time t and $\boldsymbol{\alpha}$ is the growth rate constant. This model is a first-order differential equation model that has the solution $N = N_0 * e^{\alpha t}$ where N_0 is the initial condition. For numerical solutions, MATLAB's ODE solvers can be used. The equation can be used to model population as well.

Example 8-1. For $\frac{dN}{dt} = 0.1 * N$, where N represents the number of bacteria cells, write code to plot a graph of the growth of the number of bacteria cells. The initial number of cells is 100, and the time is from 1 to 20. The code should plot the numerical solution by using `ode45` solver as well. Both exact and numerical solutions should be shown for the same figure.

Solution 8-1. The following code can be used to accomplish the given task.

I. Turk, *Practical MATLAB*, https://doi.org/10.1007/978-1-4842-5281-9_8

Example8p1.m

```
%Example8p1
%This code solves first-order ODE
%which models Bacterial Growth
clear all;close all;
time=0:0.1:20;
Alpha=0.1;% Grow Rate
N_0=100;%initial condition
N=N_0*exp(Alpha*time); %Exact Solution
[x,y]=ode45(@MyFunc,[0,20],100);
plot(time,N,x,y,'ro');
xlabel('time:t');title('Population Changes');
ylabel('Population');grid on;
function Out=MyFunc(x,y)
Alpha=0.1;
Out=Alpha*y;
end
```

Once the code is executed, the output shown in Figure 8-1 is obtained.

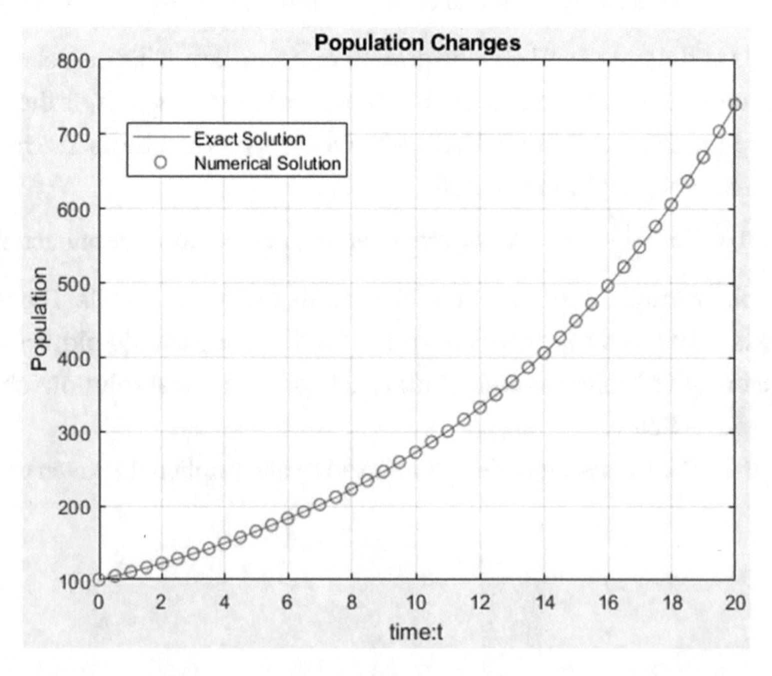

Figure 8-1. *Analytical and numerical solutions*

In Figure 8-1 the red circles shows the numerical solution and the blue line represents the exact solution. They are very close to each other, which gives us some idea about how the change occurs.

One might question the reason for finding numerical solutions as shown here. In some cases, you might not know the exact or analytical solution of the question. There might be even no solution for a question. In such cases, you might need to know what the solution looks like. In this regard, you can use numerical methods or solutions to find an answer to your question up to a certain level of accuracy. Due to the imprecision of the solution, numerical solutions are sometimes called approximated solutions.

The differential equation,

$$\frac{dN}{dt} = rN\left(1 - \frac{N}{K}\right) \quad (1)$$

is called *logistic growth* or a *logistic equation* where K is the environmental capacity and r is the rate of increase. This model can be used for bacterial growth, or a population model. The solution of this logistic equation is $N(t) = \frac{KN_0}{N_0 + e^{-rt}(K - N_0)}$.

Example 8-2. Write code to solve the logistic equation with $K = 100$, $r = 0.1$, $N_0 = 2$, and time from 1 to 100. The code should animate the analytical solution. Then, both numerical and analytical solutions should be plotted on the same figure.

Solution 8-2. The following code can be used to accomplish the given tasks.

Example8p2.m

```
%Example8p2
%This code solves logistic equation
clear all;close all;
K=500;% environmental capacity
N0=2;% Initial population size
r=0.1;% growth rate
time=linspace(1,100,100);
%N=zeros(1,length(time));
% analytic (exact) solution
```

```
for i=1:length(time) % Animation part
   N(i)=K/(1+(K/N0-1)*exp(-r*time(i)));
   plot(N,'b');
   drawnow
end
[Tode,Node]=ode45(@lge,[1 100],2);
plot(time,N,Tode,Node,'r');
xlabel('Time');title('Population Changes');
ylabel('Population');grid on;
legend('Analytical Solution',...
    'Numerical Solution');
function dN=lge(T,N)
r=0.1;
K=500;
dN=r.*N.*(1-N/K);
end
```

Once the code is executed, we obtain the output shown in Figure 8-2.

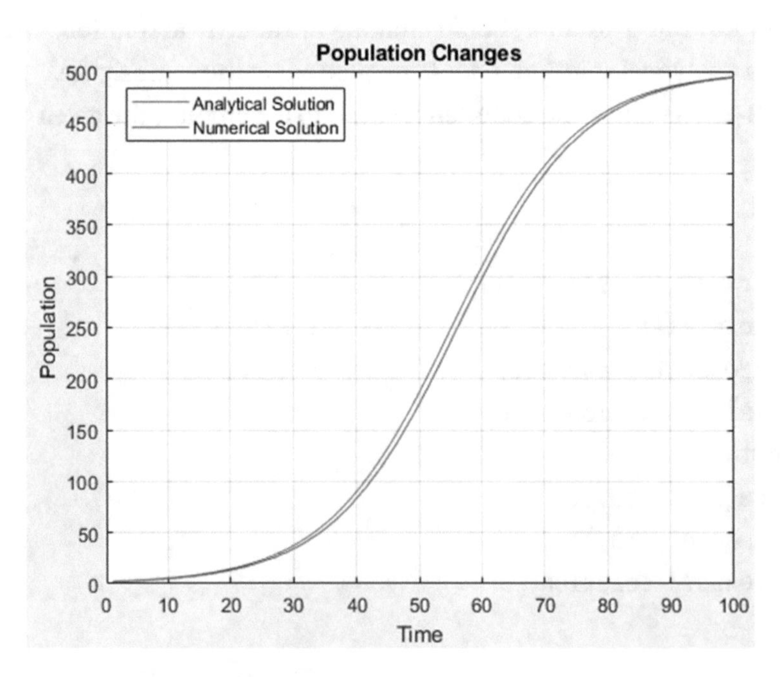

Figure 8-2. *Graph of logistic growth*

Host–Parasitoid Models

A parasitoid is an organism that lives very close to its host to benefit from it. Parasitoids and their hosts generally take advantage of each other. The *Nicholson-Bailey model* is a discrete time model used to simulate a host-parasitoid relationship.

$$\begin{cases} H_{t+1} = rH_t \exp(-aP_t) \\ P_{t+1} = eH_t \left(1 - \exp(-aP_t)\right) \end{cases} \quad (2)$$

In Equation 2, H_t stands for the density of host species and P_t stands for density of parasitoid organisms in generation t. Also r is the number of eggs laid by the host, e is the number of eggs laid by the parasitoid, and a is proportionality constant.

Example 8-3. Write code to simulate the Nicholson-Bailey model. The initial populations for host and parasitoid are 22 and 10, respectively. The number of generation is 25. The coefficients are $e = 1$, $a = 0.0069$, and $r = 2$.

Solution 8-3. The following code can be used to accomplish the given task.

Example8p3.m

```
%Example8p3
%This code solves Nicholson-Bailey model
clear all;close all;
e=1; a=0.069; r=2;
Ngens=25;  %number of generations
time=1:1:Ngens;
H=zeros(1,Ngens); P=zeros(1,Ngens);
P(1)=10; H(1)=22;%initial conditions
for t=1:Ngens-1
    H(t+1)=r*H(t)*exp(- a*P(t));
    P(t+1)=e*H(t)*(1-exp(-a*P(t)));
end
plot(time,H,'r',time,P,'b');
legend('Host','Parasitoid');
xlabel('Generation t');ylabel('Population');
title('Nicholson-Bailey Model');grid on;
```

Once the code is executed, we obtain the output shown in Figure 8-3.

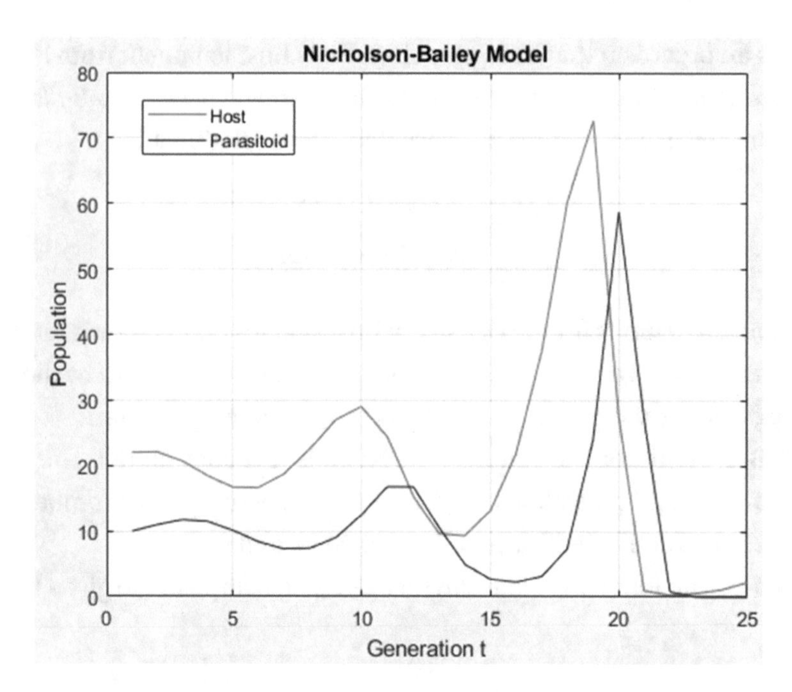

Figure 8-3. *Graph of Nicholson-Bailey model*

Bioinformatics

Bioinformatics is a broad interdisciplinary field. It develops methods that include algorithms and software to understand biological data. Dealing with the full content of this topic is outside of the scope of this book. This section presents a few applications related to genes from molecular biology.

Genome Sequencing

Genes are made up of deoxyribonucleic acid (DNA). DNA has four different bases (nucleotides): adenine (A), thymine (T), guanine (G), and cytosine (C). A is paired with T, G is paired with C, or vice versa on a strand of DNA. Genome sequencing is used to investigate the order of nucleotides in DNA.

Example 8-4. A sequence of DNA is given as TCAGGGAATTCCTACTTTTGTAT TCGCCAG. Write a code to transcribe that sequence to RNA.

Solution 8-4. If you have Bioinformatics Toolbox installed in MATLAB, then by just using built-in functions, the code can easily be transcribed into `rna`. In this solution, though, we will write all the code required to get the solution.

Example8p4.m

```
%Example8p4
%This code transcribes DNA code to RNA
Seq='TCAGGGAATTCCTACTTTTGTATTCGCCAG';
Result = MyDNA2RNA(Seq);
fprintf('DNA : %s\n',Seq')
fprintf('RNA : %s\n',Result')
```

Here, the sequence `seq` is passed to the defined function `MyDNA2RNA`. That function is saved in a different file so it can be used in different examples as well. The defined function does the transcribing. After obtaining the result, the code prints both the original code and the transcribed code.

MyDNA2RNA.m

```
function [Result]=MyDNA2RNA(Seq)
L=length(Seq);
Result ='';
for i=1:L
    switch Seq(i)
        case 'A'
            Result(i) = 'U';
        case 'T'
            Result(i) = 'A';
        case 'G'
            Result(i) = 'C';
        case 'C'
            Result(i) = 'G';
        otherwise
            disp('Code has WRONG Letter');
    end
end
end
```

Once the code is executed, we obtain the following output.

```
>> Example8p4
DNA : TCAGGGAATT C CTA C TTT T GTATTCGCCAG
RNA : AGUCCCUUAAGGAUGAAAACAUAAGCGGUC
>>
```

Example 8-5. A sequence of a DNA is given as TCAACCGGTCGTAGCTAC. Write code to transcribe it to RNA. Then, by using a genetic code table (Figure 8-4), translate it to a sequence of amino acids. Finally, the code should display the final sequence on the screen.

Solution 8-5. In the genetic code table, the nucleotides are separated, or grouped as three by three. Every group is called a *codon*. For each of these codons, there is a corresponding amino acid except UAA, UAG, and UGA. These three codons are known as *nonsense codons*.

FIRST LETTER	SECOND LETTER: U		C		A		G		THIRD LETTER
U	UUU	Phe	UCU	Ser	UAU	Tyr	UGU	Cys	U
U	UUC	Phe	UCC	Ser	UAC	Tyr	UGC	Cys	C
U	UUA	Leu	UCA	Ser	UAA	NONE	UGA	NONE	A
U	UUG	Leu	UCG	Ser	UAG	NONE	UGG	Trp	G
C	CUU	Leu	CCU	Pro	CAU	His	CGU	Arg	U
C	CUC	Leu	CCC	Pro	CAC	His	CGC	Arg	C
C	CUA	Leu	CCA	Pro	CAA	Gln	CGA	Arg	A
C	CUG	Leu	CCG	Pro	CAG	Gln	CGG	Arg	G
A	AUU	Ile	ACU	Thr	AAU	Asn	AGU	Ser	U
A	AUC	Ile	ACC	Thr	AAC	Asn	AGC	Ser	C
A	AUA	Ile	ACA	Thr	AAA	Lys	AGA	Arg	A
A	AUG	Met	ACG	Thr	AAG	Lys	AGG	Arg	G
G	GUU	Val	GCU	Ala	GAU	Asp	GGU	Gly	U
G	GUC	Val	GCC	Ala	GAC	Asp	GGC	Gly	C
G	GUA	Val	GCA	Ala	GAA	Glu	GGA	Gly	A
G	GUG	Val	GCG	Ala	GAG	Glu	GGG	Gly	G

Figure 8-4. *Genetic code table*

All amino acids are listed in Figure 8-4 with three-letter abbreviations. As an example, AUU represents isoleucine amino acid, and it is abbreviated Ile. It can be just shown by its first letter, I, as well. Some amino acids start with the same letters, however. Different single letters are assigned to these amino acids. These single codes are taken from MATLAB's official web site. In this example, we will show each amino acid with both three letters and one letter.

Example8p5.m

```
%Example8p5
%This code transcribes DNA code to RNA
Seq='TCAGGGAATTCCTACTTTTGTATTCGCCAG';
Convert = MyDNA2RNA(Seq);
fprintf('DNA : %s\n',Seq');
fprintf('RNA : %s\n',Convert);
Codons = MyCodons(Convert);
disp('Codons :');
disp(Codons);
[Codn3,Codn1]= GeneticCode(Codons);
disp('Amino Acids with 3 codes');
disp(Codn3);
disp('Amino Acids with 1 code');
disp(Codn1);
```

In the preceding code, there are three defined functions: `MyDNA2RNA`, `MyCodons`, and `GeneticCode`. The first function is the same as in the previous question. The other two are shown in the following code.

MyCodons.m

```
function [Out] = MyCodons(Seq)
%This function takes the sequence
%and returns them as codons
L=length(Seq);
Subst=mod(L,3);
Seq2=cell(1,(L-Subst)/3);
```

```
for k=1:(L/3)
    Seq2{k} = Seq(3*k-2:3*k);
end
Out = Seq2;
end
```

GeneticCode.m

```
function [Out1,Out2] = GeneticCode(Seq)
L=length(Seq);%Below are 20 amino acids
Seq2=cell(1,L);Seq3=cell(1,L);
Phe=['UUU','UUC'];Ile=['AUU','AUC','AUA'];
Leu=['UUA','UUG','CUU','CUC','CUA','CUG'];
Met=['AUG'];NONE=['UAA','UAG','UGA'];
Val=['GUU','GUC','GUA','GUG'];
Ser=['UCU','UCC','UCA','UCG','AGU','AGC'];
Pro=['CCU','CCC','CCA','CCG'];
Thr=['ACU','ACC','ACA','ACG'];
Ala=['GCU','GCC','GCA','GCG'];
Tyr=['UAU','UAC'];His=['CAU','CAC'];
Gln=['CAA','CAG'];Asn=['AAU','AAC'];
Lys=['AAA','AAG'];Asp=['GAU','GAC'];
Glu=['GAA','GAG'];Cys=['UGU','UGC'];
Arg=['CGU','CGC','CGA','CGG','AGA','AGG'];
Gly=['GGU','GGC','GGA','GGG'];Trp=['UGG'];
for i=1:(L)
    if contains(Seq(i),Phe)
        AA='Phe';Aa='F';
    elseif contains(Ile,Seq(i))
        AA='Ile';Aa='I';
    elseif contains(Leu,Seq(i))
        AA='Leu';Aa='L';
    elseif contains(Met,Seq(i))
        AA='Met';Aa='M';
```

```
    elseif contains(NONE,Seq(i))
        AA='NON';Aa='-';
    elseif contains(Val,Seq(i))
        AA='Val';Aa='V';
    elseif contains(Ser,Seq(i))
        AA='Ser';Aa='S';
    elseif contains(Pro,Seq(i))
        AA='Pro';Aa='P';
    elseif contains(Thr,Seq(i))
        AA='Thr';Aa='T';
    elseif contains(Ala,Seq(i))
        AA='Ala';Aa='A';
    elseif contains(Tyr,Seq(i))
        AA='Tyr';Aa='Y';
    elseif contains(Gln,Seq(i))
        AA='Gln';Aa='Q';
    elseif contains(Lys,Seq(i))
        AA='Lys';Aa='K';
    elseif contains(Glu,Seq(i))
        AA='Glu';Aa='I';
    elseif contains(Arg,Seq(i))
        AA='Arg';Aa='R';
    elseif contains(Gly,Seq(i))
        AA='Gly';Aa='G';
    else
        AA='---';Aa='-';
    end
    Seq2{i}=AA;Seq3{i}=Aa;
end
Out1=Seq2;Out2=Seq3;
end
```

Once the code is run, the following output is displayed.

```
> Example8p5
DNA : TCAGGGAATTCCTACTTTTGTATTCGCCAG
RNA : AGUCCCUUAAGGAUGAAAACAUAAGCGGUC
Codons :
  Columns 1 through 8
    'AGU'    'CCC'    'UUA'    'AGG'    'AUG'    'AAA'    'ACA'    'UAA'
  Columns 9 through 10
    'GCG'    'GUC'
Amino Acids with 3 codes
  Columns 1 through 8
    'NON'    'Pro'    'Ile'    'Arg'    'Met'    'Lys'    'Thr'    'NON'
  Columns 9 through 10
    'Ala'    'Val'
Amino Acids with 1 code
    '-'    'P'    'I'    'R'    'M'    'K'    'T'    '-'    'A'    'V'
>
```

One dash or three dashes belong to the nonsense codons.

Example 8-6. From the previous example, where the sequence is TCAACCGGTCGTAGCTAC, remove the first letter, and transcribe the new sequence again. Write the obtained amino acids as one single word where each letter represents one amino acid. Explain the obtained result.

Solution 8-6. The following code can be used to accomplish the given tasks.

Example8p6.m

```
%Example8p6
%This code transcribes DNA code to RNA
Seq='TCAGGGAATTCCTACTTTTGTATTCGCCAG';
Seq2=Seq(2:end);
Convert = MyDNA2RNA(Seq2);
Codons = MyCodons(Convert);
[Codn3,Codn1]= GeneticCode(Codons);
MySeq='';
```

```
for k=1:length(Codn1)
    MySeq(k)=Codn1{k};
end
disp('Original Sequence:');disp(Seq);
disp('Updated Sequence');disp(Seq2);
disp('Obtained Amino Acids:');disp(MySeq);
```

Here, the `MyDNA2RNA`, `MyCodons`, and `GeneticCode` functions are the same functions as those used before. Once the code is run, the following output is obtained.

```
> Example8p6
Original Sequence:
TCAGGGAATTCCTACTTTTGTATTCGCCAG
Updated Sequence
CAGGGAATTCCTACTTTTGTATTCGCCAG
Obtained Amino Acids:
VL-R-KIKR
>
```

As shown, the obtained amino acids here and those obtained from the previous question are different. This indicates that, if any of the nucleobase is missed in a sequence, then a totally different sequence is obtained after transcribing.

Dot Plot

Dot plots are used to display a comparison of the given two sequences. To plot the graphic, generally a two-dimensional matrix is produced, so the obtained graph is also two-dimensional. The first given sequence is placed on the row, and the second given sequence is laid in a column of the matrix. In the matrix, 1 is assigned for a match, and 0 is assigned for a mismatch.

Example 8-7. For the given two sequences `Seq1=AATGCAATT` and `Seq2=ATTGACT`, write code to print out the dot matrix and draw a dot plot. In the graph, matches should be shown in red, and mismatches should be shown in blue.

Solution 8-7. The following code can be used to accomplish the given tasks.

Example8p7.m

```
%Example8p7
%This code creates dot matrix and dot plot
Seq1='AATGCAATT';Seq2='ATTGACT';
X=length(Seq1);Y=length(Seq2);
DotMatrix=zeros(Y,X);
for i=1:Y
    Places=strfind(Seq1,Seq2(i));
    DotMatrix(i,Places)=1;
end
disp('The Dot Matrix:');
disp(DotMatrix);figure;hAxes=gca;
imagesc(hAxes,DotMatrix);
colormap(hAxes,[0,0,1;1,0,0]);
ylabel(fliplr(Seq2));title(Seq1);
```

In the preceding code, the matching places are found for each member of the Seq2 variable using the `strfind` command. Then, 1 is assigned to these places in `DotMatrix`. After creating the matrix, the figure is created using the `figure` function. The gca command handles the axis of the current figure, and the figure is displayed with scaled colors set using the `imagesc` function. Match and mismatch color assignment is achieved using the `colormap` command for the current figure. The order of the Seq2 variable is reversed using the `fliplr` command, and it is put on the *y* axis where the Seq2 variable is placed as a title.

Once the code is executed, we obtain the following output.

```
> Example8p7
The Dot Matrix:
     1     1     0     0     0     1     1     0     0
     0     0     1     0     0     0     0     1     1
     0     0     1     0     0     0     0     1     1
     0     0     0     1     0     0     0     0     0
     1     1     0     0     0     1     1     0     0
     0     0     0     0     1     0     0     0     0
     0     0     1     0     0     0     0     1     1

>
```

The image shown in Figure 8-5 is displayed.

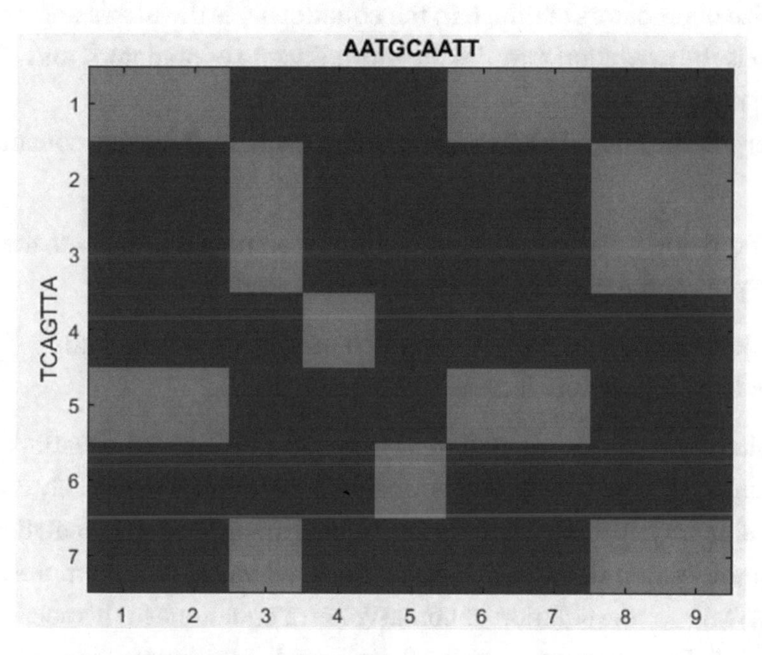

Figure 8-5. *Dot plot image of dot matrix*

Predator–Prey Models

Predators are basically the organisms that eat other organisms to survive. Preys are also the organisms that are eaten by the predators. Some examples between predators and preys can be given as lions with zebras, tigers with buffalos, and sharks with fishes.

Model with Two Species

The most famous model used to describe the relationship between predators and their prey is the *Lotka-Volterra equations.*

$$\begin{cases} \dfrac{dA}{dt} = \gamma A - \alpha AB; \\ \dfrac{dB}{dt} = -\delta B + \beta AB. \end{cases} \tag{3}$$

A(t) represents the population size of prey at time *t*, and *B(t)* represents the population size of predators at time *t*. In the equations, γ is the increasing rate of prey population, α is the predation rate, δ is the mortality rate of predators, and β is the reproduction rate of predators.

If we interpret the Lotka-Volterra equations shown, the following conclusions can be reached.

- In Equation 3, if there are no predators, assuming that $B(t) = 0$, then the prey population is exponentially increased.
- In Equation 3, if no prey exists, which means $A(t) = 0$, then the predator population decreases exponentially.

Lotka-Volterra equations are nonlinear, first-order differential equations. For such types of nonlinear differential equations, analytical solutions are unlikely due to the nature of nonlinearity. Although some reports have mentioned that analytical solutions exist under some assumptions for these equations, we will benefit from the numerical solvers of MATLAB for the solution of Lotka-Volterra equations in this section.

Example 8-8. For the given Equation 3, write code to simulate the population sizes for predators and prey with respect to time. The initial populations are 10 and 100 for predators and prey, respectively. Time span will be from 0 to 50. Other coefficients are given as $\gamma = 0.4$, $\alpha = 0.4$, $\delta = 2.0$, and $\beta = 0.1$. Explain the graphic obtained. The figure name should have the title Predator-Prey Model, and there should be no menu bar on the figure.

Solution 8-8. The following code can be used to accomplish the given tasks.

Example8p8.m

```
%Example8p8
%This code solves Lotka-Volterra equations
y0 = [100;10];%initial populations
[t,x] = ode45(@LVfunc,[0 50],y0);
A = x(:,1);%Prey
B = x(:,2);%Predator
figure('Name','Predator-Prey Model',...
    'NumberTitle','off','MenuBar', 'none');
plot(t,A,'-o',t,B,'r-');grid on;
legend('Preys','Predators');
```

```
title('Population Along With Time')
xlabel('Time');ylabel('Population');
function dxdt = LVfunc(t,x)
gama = 0.4; alpha = 0.4;
delta = 2.0; beta = 0.1;dxdt=[0;0];
dxdt(1)=gama*x(1)-alpha*x(1)*x(2);%prey
dxdt(2)=-delta*x(2)+beta*x(1)*x(2);%predators
end
```

Once the code is run, the output shown in Figure 8-6 is displayed.

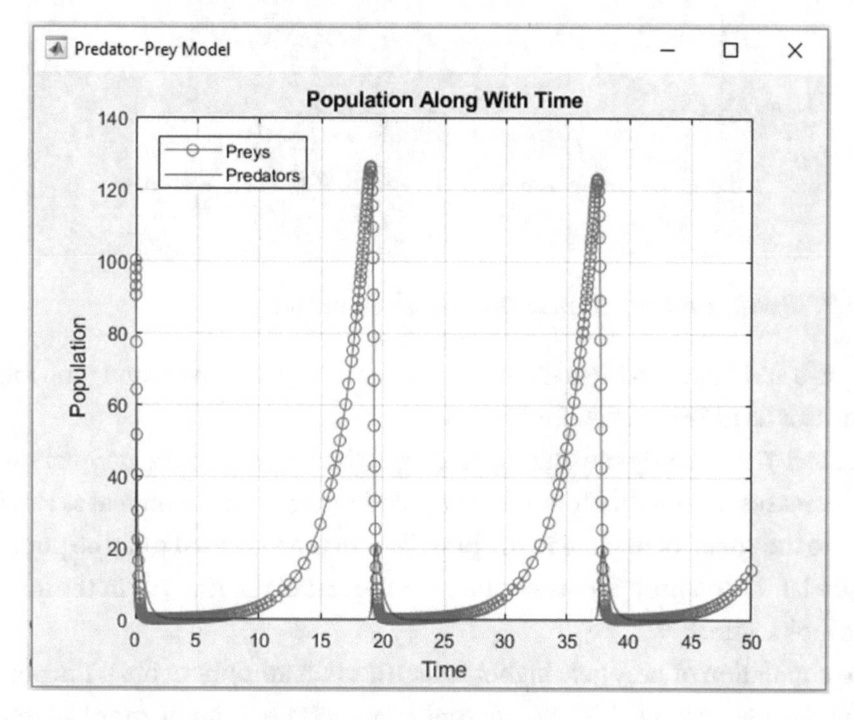

Figure 8-6. *Predator–prey population model*

In Figure 8-6, we see that population starts with 100 and 10 for predators and prey, respectively. To see clearly what happens at the very beginning horizontally, in the ninth row of the code, we can use the `semilog` function instead of the `plot` function. By changing just these two functions, we get the graph shown in Figure 8-7.

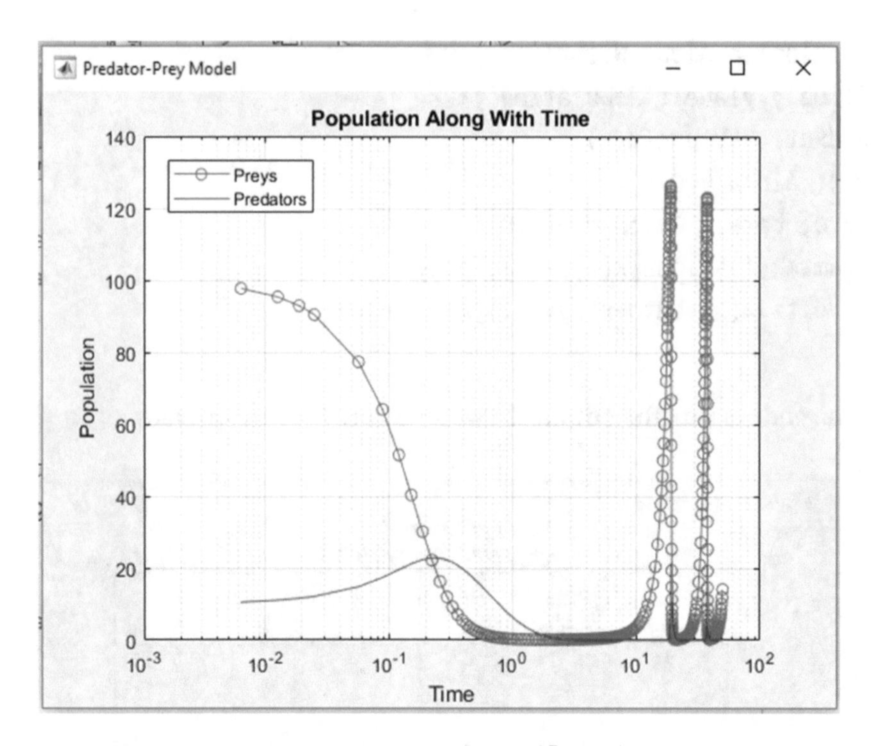

Figure 8-7. *Predator-prey populations with* `semilog`

Figure 8-6 and Figure 8-7 use the same *x* and *y* values. In the second plot, logarithmic scale values for base 10 are used for the *x* axis.

In Figure 8-7, we clearly see that starting from the beginning, the population of the prey decreases exponentially and the population of predators increases slowly, compared to the speed of decline in the prey. Before time is equal to 1, the populations for both predators and prey become equal. For the rest of the time, or in the long run, we can take a look Figure 8-6.

As the population of prey gets higher, after a certain amount of time, the population of predators increases as well. The predators eat prey faster than the prey can reproduce. As a result of, the population of prey begins to fall. The food supply thus diminishes for predators, so their population begins to fall as well. After some time, the whole scenario starts repeating itself, and the populations behave predictably with respect to time.

Model with Three Species

There exists a modified Lotka-Volterra model for three species. This model has a system of three equations.

$$\frac{dA}{dt} = \gamma A - \alpha AB; \tag{4}$$

$$\frac{dB}{dt} = -\delta B + \beta AB - eBC; \tag{5}$$

$$\frac{dC}{dt} = -fC + gBC. \tag{6}$$

Here C is another type of organism besides A and B, which were introduced in the previous section. C can hunt the predator B when there is an interaction. The population of organism C can go down with the absence of predators. According to Equations 4, 5, and 6, there is no interaction between organisms A and C.

Example 8-9. For the given Equations 4, 5, and 6, write code to simulate the population sizes for predators, prey, and the third organism with respect to time. The initial population is 100 for all three organisms. Time span will be from 0 to 60. Other coefficients are given as $\gamma = 0.4$, $\alpha = 0.4$, $\delta = 2.0$, $\beta = 0.2$, $e = 0.2$, $f = 0.1$, and $g = 0.15$.

Solution 8-9. The following code can be used to accomplish the given tasks.

Example8p9.m

```
%Example8p9
%Lotka-Volterra eqns with 3 species
clear all;close all;
y0 = [100;100;100];%initial populations
[t,x] = ode45(@LVfunc2,[0 60],y0);
A = x(:,1);%Prey
B = x(:,2);%Predator
C = x(:,3);%3rd Organism
figure('Name','Predator-Prey Model',...
    'NumberTitle','off','MenuBar', 'none');
plot(t,A,'-o',t,B,'r-',t,C,'gx');grid on;
legend('Preys','Predators','3rd organism');
title('Population Along With Time')
xlabel('Time');ylabel('Population');
```

LVfunc2.m

```
function dxdt = LVfunc2(t,x)
dxdt=[0;0;0]; gama = 0.4; alpha = 0.4;
delta = 2.0; beta = 0.2;
e=0.2;f=0.1;g=0.15;
dxdt(1)=gama*x(1)-alpha*x(1)*x(2);%prey
dxdt(2)=-delta*x(2)+beta*x(1)*x(2)-...
    e*x(2)*x(3);%predators
dxdt(3)=f*x(2)*x(3)-g*x(3);%3rd organism
end
```

Once the code is run, the output in Figure 8-8 is displayed.

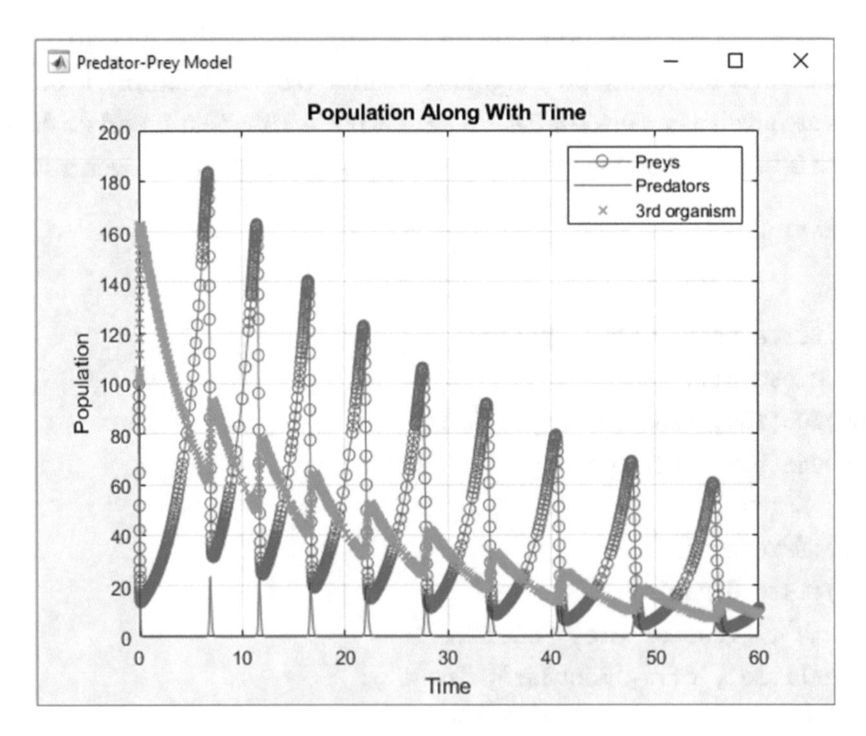

Figure 8-8. *Populations of three species*

In Figure 8-8, we can see the long run of populations. By just looking at this output, we can say that populations of prey and the third organism decrease over time. This picture is misleading, however. If we increase the final time from 60 to 100, we see that the bars for each organism stabilize, and they behave predictably after a certain amount of time. We can see this scenario in Figure 8-9. where the final time is entered as 200.

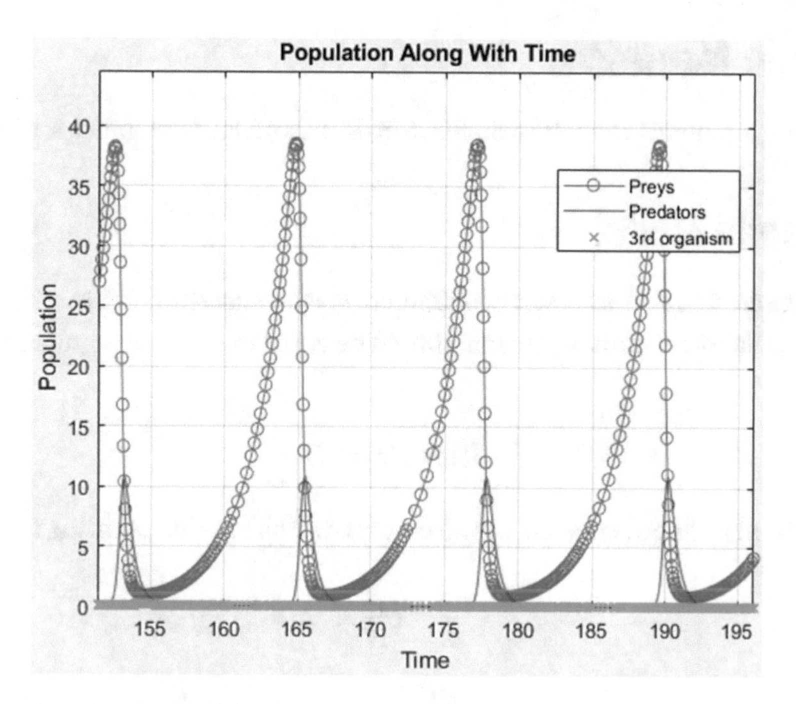

***Figure 8-9.** Populations of three species at the outset*

For the very beginning of the dynamics, we can take a look at the output in Figure 8-10.

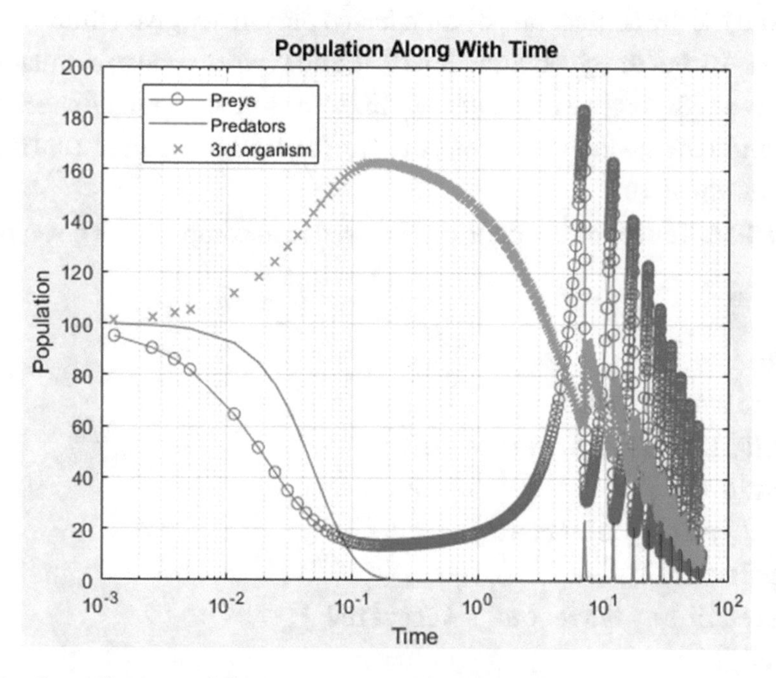

***Figure 8-10.** Population of three species at the outset*

Epidemic Models

In this section, epidemic models of SI, SIS, SIR, SIRS, and HIV dynamics are presented.

SI Epidemic Model

In this epidemic model, there exist two different stages, susceptible (S), and infected (I). The total population of both organisms should be equal to the total population size, *N*. In other words,

$$S(t)+I(t)=N. \tag{7}$$

The SI model without considering births and deaths has the following form:

$$\frac{dS}{dt}=-\frac{\beta}{N}SI, \tag{8}$$

$$\frac{dI}{dt}=\frac{\beta}{N}SI. \tag{9}$$

Here, $\boldsymbol{\frac{\beta}{N}SI}$ represents the number of infections due to infected individuals at a given time, and $\boldsymbol{\beta}$ is the contact rate by an infected person at a given time.

Example 8-10. For the given Equations 7, 8, and 9, write code to simulate the population sizes of SI stages with respect to time. The initial populations are 9 for susceptible, and 1 for the infected group, making the total population 10. Time span will be from 0 to 10, and $\beta = 0.9$.

Solution 8-10. The following code can be used to accomplish the given tasks.

Example8p10.m

```
%Example8p10
%This code solves SI model
[t,x]=ode45(@SI,[0,10],[9,1]);
plot(t,x);grid on;
xlabel('time');title('SI Model');
ylabel('population');
legend('Susceptible','Infected','Recovered');
function Si=SI(t,x)
beta=.9;N=10;
```

```
%x(1)=susceptible group
Si(1,:)=-(beta/N)*x(1)*x(2);
%x(2)=exposed by infection
Si(2,:)=(beta/N)*x(1)*x(2);
end
```

Once the code is run, the output shown in Figure 8-11 is displayed.

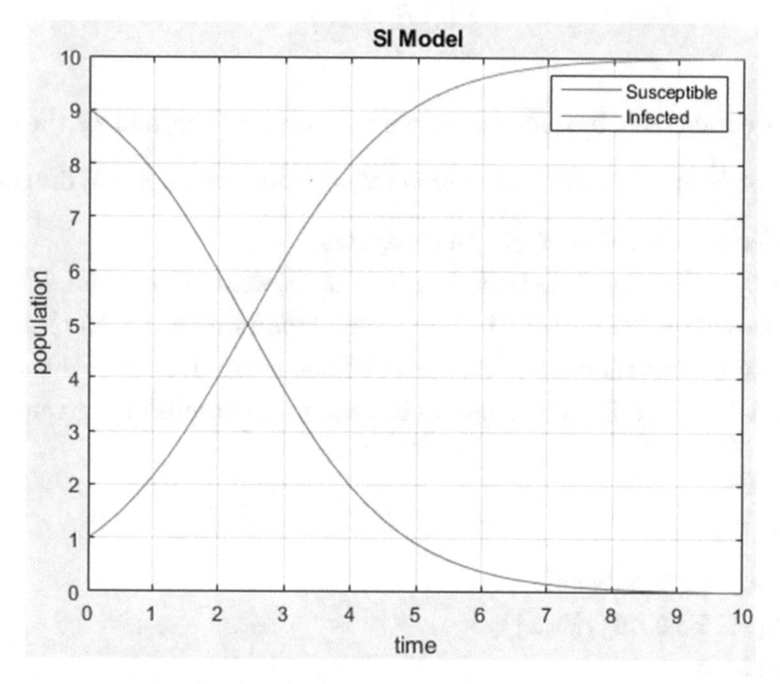

Figure 8-11. *Dynamics of SI model*

As shown in Figure 8-11, the infected population starts with 1 person, and at the end, all populations have become infected. Conversely, the susceptible population starts with 9 persons and is reduced to 0 at the end of 10 days as everybody went from the S to I stage. In this model, everyone becomes infected when time goes to infinity. Therefore, it can be concluded that this model is more suitable for highly infectious diseases.

SIS Epidemic Model

In this epidemic model, it looks like as if there are three stages: susceptible (S) stage, infected (I) stage, and susceptible (S) stage again. The last stage is actually the first stage. Therefore, actually only two stages exist in the SIS model. All infected organisms become susceptible again. This type of model is more applicable to the diseases that commonly have repeat infections in a cycle.

The total population of the all organisms should be equal to total population size, *N*. The SIS model without considering births and deaths has the following form:

$$\frac{dS}{dt} = -\frac{\beta}{N}SI + \gamma I \tag{10}$$

$$\frac{dI}{dt} = \frac{\beta}{N}SI - \gamma I \tag{11}$$

where β is the contact rate by infected people at a point in time, and γ is the recovery rate.

The ratio of $\frac{\beta}{\gamma}$ is called the *basic reproduction number* R_0. If $R_0 > 1$, the disease becomes endemic, otherwise ($R_0 \le 1$) it dissipates.

Example 8-11. For the given Equations 10 and 11, write code to simulate the population sizes of the SIS model. The initial populations are 9 and 1 for S and I states, respectively. Time span is from 0 to 50. The coefficients are given as $\gamma = 1.0$ and $\beta = 0.9$.

Solution 8-11. The following code can be used to accomplish the given tasks.

Example8p11.m

```
%Example8p11
%This code solves SIS model
[t,x]=ode45(@SIS,[0,10],[9,1]);
plot(t,x);grid on;
xlabel('time');
title('SIS Model with ${R}_{0} \ge 1$',...
    'FontAngle','italic','Interpreter',...
    'Latex');
ylabel('population');
legend('Susceptible','Infected');
function Si=SIS(t,x)
beta=.9;N=10;
gamma=1.0;%Recovery rate
%x(1)=susceptible group
```

```
Si(1,:)=-(beta/N)*x(1)*x(2)+gamma*x(2);
%x(2)=exposed by infection
Si(2,:)=(beta/N)*x(1)*x(2)-gamma*x(2);
end
```

In the preceding code, to print $R_0 \leq 1$ as the title, a LaTeX interpreter is used. Once the code is run, the output shown in Figure 8-12 is the result.

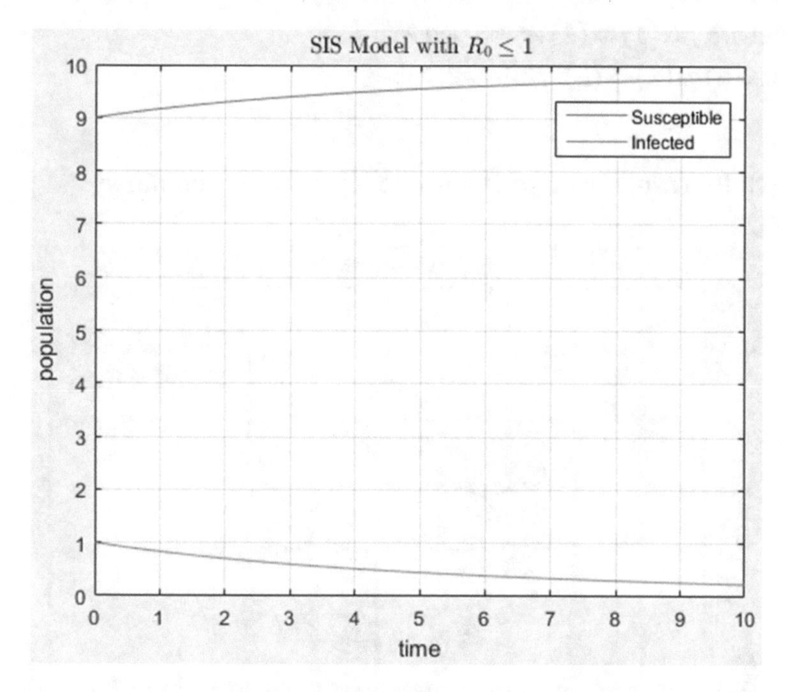

Figure 8-12. *Dynamics of SIS model when $R_0 \leq 1$*

It is obvious that the disease fades out over time.

Example 8-12. Simulate the SIS dynamics with the same population sizes and time span used for Example 8-11. The coefficients for this example are given as $\gamma = 0.1$, and $\beta = 0.9$ to make $R_0 > 1$.

Solution 8-12. The following code can be used to accomplish the given tasks.

Example8p12.m

```
%Example8p12
%This code solves SIS model
[t,x]=ode45(@SIS,[0,10],[9,1]);
plot(t,x);grid on;xlabel('time');
```

```
title('SIS Model with ${R}_{0} \ge 1$',...
   'Interpreter','Latex');
ylabel('population');
legend('Susceptible','Infected');
function Si=SIS(t,x)
beta=.9;N=10;
gamma=0.1;%Recovery rate
Si(1,:)=-(beta/N)*x(1)*x(2)+gamma*x(2);
Si(2,:)=(beta/N)*x(1)*x(2)-gamma*x(2);
end
```

Once the code is run, the output shown in Figure 8-13 is obtained.

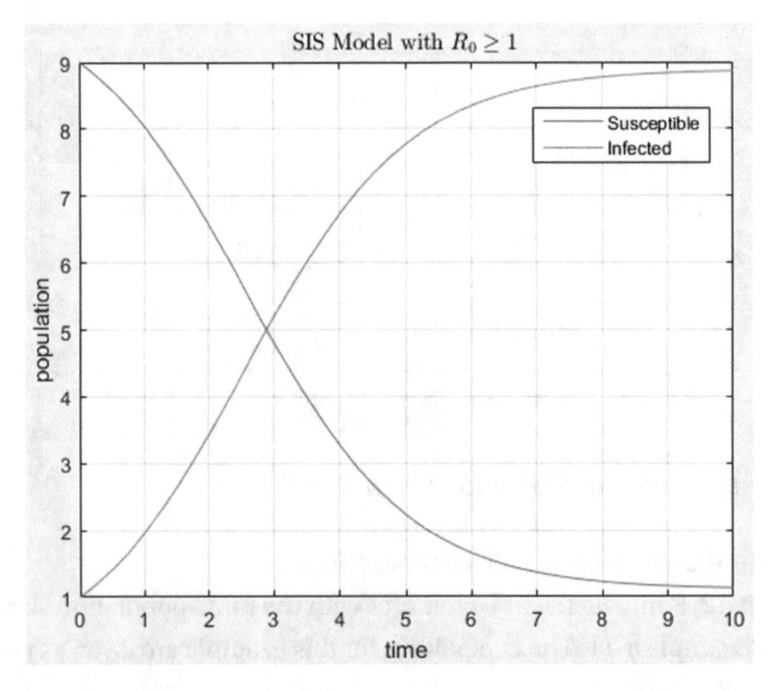

***Figure 8-13.** Dynamics of SIS model when $R_0 > 1$*

As seen in Figure 8-13, when $R_0 > 1$, the disease spreads to the whole population.

SIR Epidemic Model

In this epidemic model, there are three different stages: susceptible (S), infected (I), and recovered (R). The total population of the all organisms should be equal to the total population size, *N*. In other words,

$$S(t)+I(t)+R(t)=N. \tag{12}$$

The SIR model without considering births and deaths has the following form:

$$\frac{dS}{dt}=-\frac{\beta}{N}SI \tag{13}$$

$$\frac{dI}{dt}=\frac{\beta}{N}SI-\gamma I \tag{14}$$

$$\frac{dR}{dt}=\gamma I \tag{15}$$

where β is the contact rate by an infected people at a point in time and γ is the recovery rate. $R_0=\frac{\beta}{\gamma}$ is the basic reproduction number.

Example 8-13. For the given Equations 13, 14, and 15, write code to simulate the population sizes for all S, I, and R stages over time. The initial populations are 9, 1, and 0 for S, I, and R populations, respectively. Time span is from 0 to 50. Other coefficients are given as $\gamma = 0.1$ and $\beta = 0.9$.

Solution 8-13. The following codes can be used to accomplish the given tasks.

Example8p13.m

```
%Example8p13
%This code solves SIR model
[t,x]=ode45(@SIR,[0,50],[9,1,0]);
%semilogx(t,x);
plot(t,x);
xlabel('time');ylabel('population');
title('SIR Model');grid on;
legend('Susceptible','Infected','Recovered');
function Sir=SIR(t,x)
```

```
beta=.9;%contact rate by people in population i
gamma=.1;%Recovery rate
N=10;%N=Total number of population
%dS/dt=Sir(1),x(1)=susceptible people
ds=-(beta/N)*x(1)*x(2);
%dI/dt=Sir(2),x(2)=exposed by infection
di=(beta/N)*x(1)*x(2)-gamma*x(2);
%dr/dt=sir(3),x(3)=Recovered from infection
dr=gamma*x(2);
Sir=[ds;di;dr];
end
```

Once the code is run, the output displayed in Figure 8-14 is obtained.

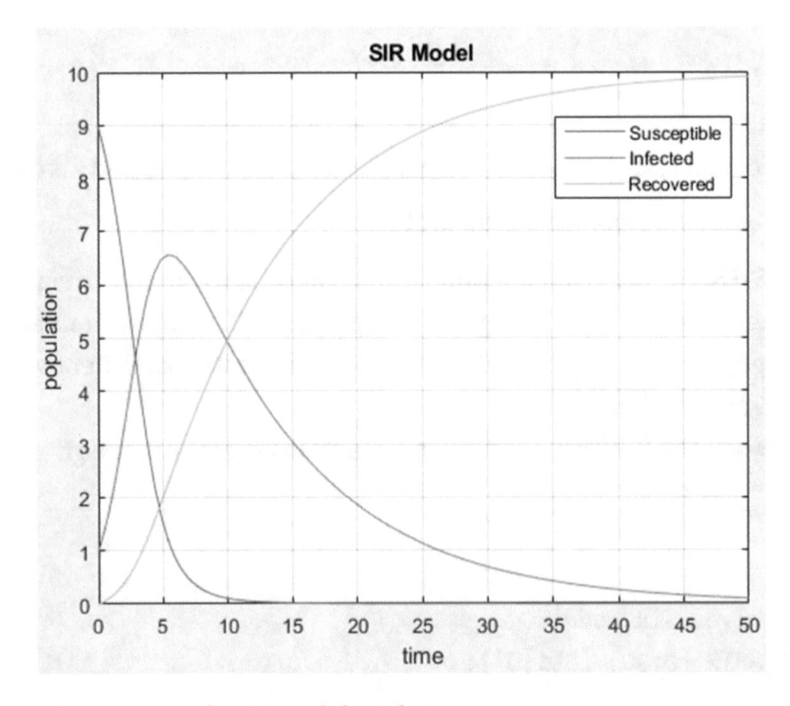

***Figure 8-14.** Dynamics of SIR model with $R_0 > 1$*

In this outcome, R_0 is greater than 1. If we use $\gamma = 1.0$ and keep the rest of the parameters in this example the same, the graphic shown in Figure 8-15 is the output when the reproduction number is equal to or less than 1.

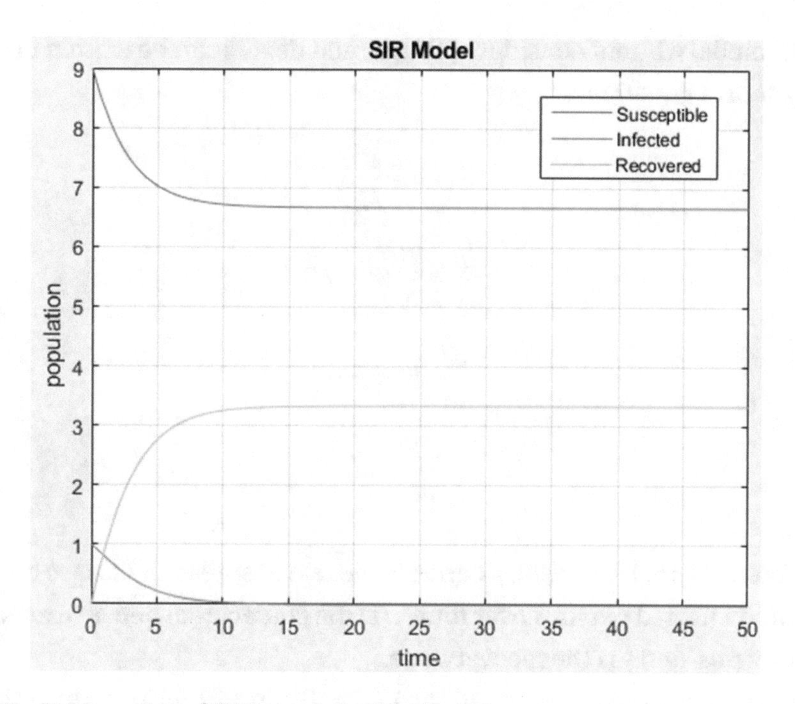

Figure 8-15. *Dynamics of SIR model with $R_0 \leq 1$*

SEIR Epidemic Model

In this epidemic model, illustrated in Figure 8-16, the exposed stage (E) is added to the other known S, I, and R stages.

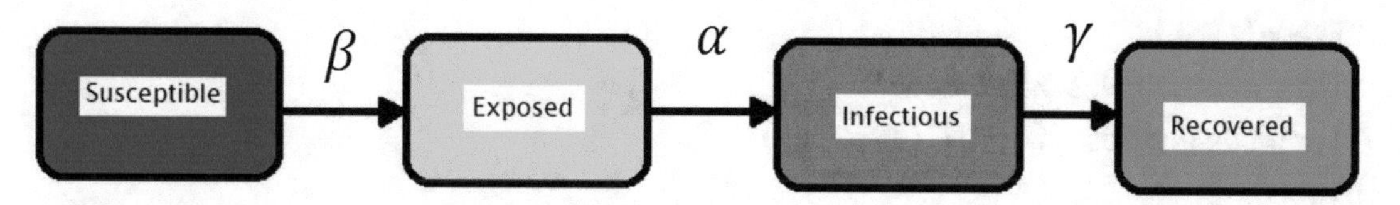

Figure 8-16. *Stages of the SEIR model*

The total population of the all organisms should be equal to the total population size, *N*. In other words,

$$S(t)+E(t)+I(t)+R(t)=N. \tag{16}$$

The SEIR model without considering births and deaths can be written in the following system of equations.

$$\frac{dS}{dt} = -\frac{\beta}{N} SI \tag{17}$$

$$\frac{dE}{dt} = \frac{\beta}{N} SI - \alpha E \tag{18}$$

$$\frac{dI}{dt} = \alpha E - \gamma I \tag{19}$$

$$\frac{dR}{dt} = \gamma I \tag{20}$$

In Equations 17, 18, 19, and 20, β controls the rate of spread of disease between susceptible and infected persons over time, α is the contact rate between exposed and infected organisms, and γ is the recovery rate.

Example 8-14. For the given Equations 17, 18, 19, and 20, write code to simulate the population sizes for all populations with respect to time. The initial populations are 17, 2, 1, and 0 for the S, E, I, and R groups, respectively. Time span is from 0 to 50. Other coefficients are $\beta = 0.8$, $\alpha = 0.4$, and $\gamma = 0.3$.

Solution 8-14. The following codes can be used to accomplish the given tasks.

Example8p14.m

```
%Example8p14
%This code solves SEIR model
[t,x]=ode45(@Seir,[0,50],[17,2,1,0]);
plot(t,x);
xlabel('time');ylabel('population');
title('SEIR Model');grid on;
legend('Susceptible',...
    'Exposed','Infected','Recovered');
function SEIR=Seir(t,x)
N=10;beta=.8;delta=0.4;gamma=1;
SEIR(1,:)=-(beta/N)*x(1)*x(3);
```

```
SEIR(2,:)=(beta/N)*x(1)*x(3)-delta*x(2);
SEIR(3,:)=delta*x(2)-gamma*x(3);
SEIR(4,:)=gamma*x(3);
end
```

Once the code is run, the output shown in Figure 8-17 is displayed.

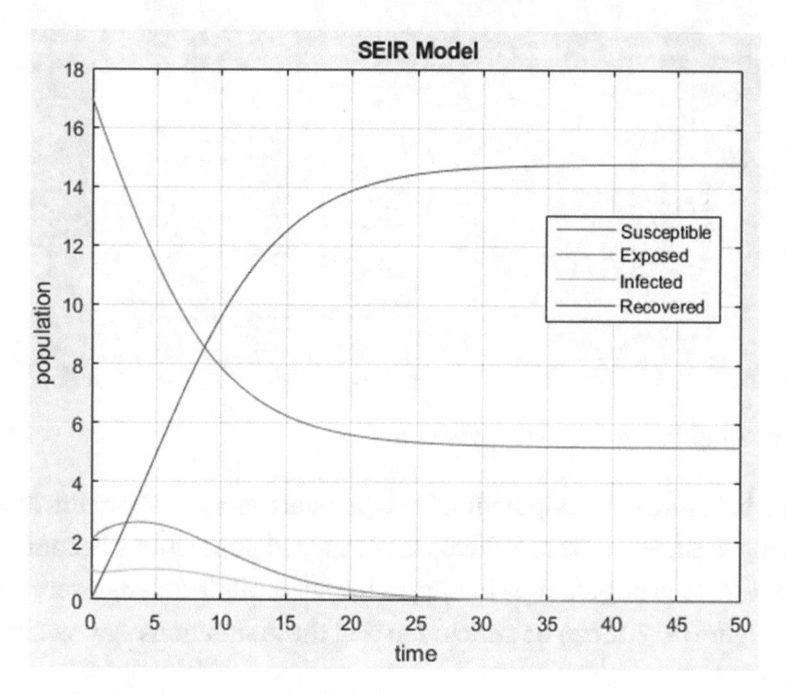

Figure 8-17. *SEIR model*

To see the very beginning of the dynamics clearly, we can substitute the `semilogx` command with the `plot` function to get the result shown in Figure 8-18.

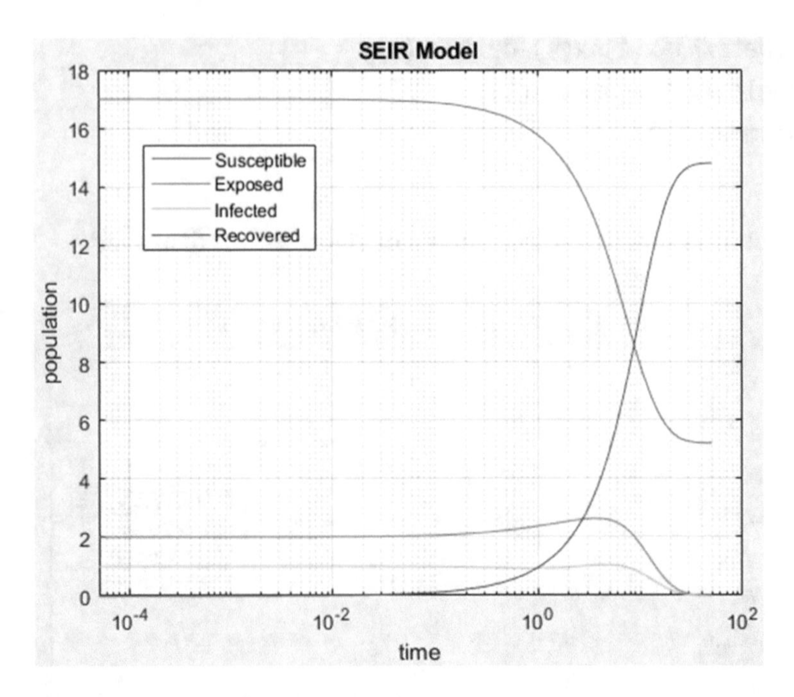

Figure 8-18. *Beginning of SEIR dynamics*

As is clear in Figure 8-18, populations of types start varying before the time is equal to 1. In the long term, we see that infected and exposed populations become almost zero and the rest of the populations belong to the susceptible and recovery states. Considering Figure 8-17, it can be concluded that these situations stay steady when time goes to infinity.

One important point we need to keep in mind is that these outcomes can be changed by just changing the coefficients of the dynamics. If we add mortality or births to the equations, then we could encounter totally a different scenario from what we saw in the previous examples.

Cellular Dynamics of HIV

Human immunodeficiency virus (HIV) targets and alters the immune system and ultimately leads to acquired immune deficiency syndrome (AIDS), which is often fatal. Researchers and scientists have been trying for decades to find solutions to protect people from HIV and AIDS.HIV attacks a specific type of cell in the body, the CD4 helper cell, or T cell. The basic model for T cell and virus dynamics has a system of three equations:

$$\frac{dx}{dt} = \gamma - d_x x - \beta x v, \tag{21}$$

$$\frac{dy}{dt} = \beta x v - d_y y, \tag{22}$$

$$\frac{dv}{dt} = ky - d_v v - \beta x v \tag{23}$$

In this system of equations, $\frac{dx}{dt}$ represents the number of uninfected cells changing with respect to time, $\frac{dy}{dt}$ represents the number of infected cells changing with respect to time, and $\frac{dv}{dt}$ represents the number of free virus particles changing with respect to time where $x_0 > 0$, $v_0 > 0$, and $y_0 = 0$ should hold for the initial conditions.

For the model of Equations 21, 22, and 23, the basic reproduction number R_0 is

$$R_0 = \frac{k \frac{\beta\gamma}{d_x}}{d_y \left(d_v + \frac{\beta\gamma}{d_x} \right)} = \frac{N\beta\gamma}{d_x d_v + \beta\gamma} \tag{24}$$

where γ is the rate of uninfected cells produced by the immune system, and $\beta x v$ is the rate of infected cells as a result of the crossing of free virus and uninfected cells. N is the number of free virus particles, which should be much greater than 1. d_x, d_y, and d_v are the death rates of uninfected cells, infected cells, and virus particles, respectively. Time is premeasured in days. k is the rate of production of free viral particles from an infected cell.

Example 8-15. For Equations 21, 22, and 23, write code to simulate the population sizes for uninfected cells, infected cells, and virus particles with respect to time. The initial populations are 10^6, 0, and 2 for uninfected cells, infected cells, and virus particles, respectively. Time span will be from 0 to 50. Other coefficients are $\gamma = 10^5$, $k = 200$, $\beta = 3 * 10^{-7}$, $d_x = 0.12$, $d_y = 0.5$, and $d_v = 4$.

Solution 8-15. The following code shows the core function used inside the main code.

Hiv.m

```
function HIV=Hiv(t,x)
dx=0.12;dy=0.5;dv=4;
beta=3e-7;k=1e+2;gamma=1e+5;
HIV(1,:)=gamma -dx*x(1)-beta*x(1)*x(3);
HIV(2,:)=beta*x(1)*x(3) -dy*x(2);
HIV(3,:)=k*x(2)-dv*x(3)-beta*x(1)*x(3);
end
```

Most of the values of the variables in the preceding code are taken from *An Introduction to Mathematical Biology* by Linda J. S. Allen (Pearson Education, 2007, 278). The main part of the code can be written as follows.

Example8p15.m

```
%Example8p15
%This code solves HIV model
clear all;close all;
[t,x]=ode45(@Hiv,[0,50],[1e+6,0,2]);
plot(t,x(:,1));
xlabel('Time in days');ylabel('Population');
title('Uninfected Cells');grid on;
figure;plot(t,x(:,2))
xlabel('Time in days');ylabel('Population');
title('Infected Cells');grid on;
figure;plot(t,x(:,3))
xlabel('Time in days');ylabel('Population');
title('Virus Particles');grid on;
```

In the preceding code, `subplot` can be used instead of a `plot` function. To display the graphics clearly, three `plot` functions are used instead. Once the `Example8p15.m` file is run, we obtained the output shown in Figure 8-19 through 8-21.

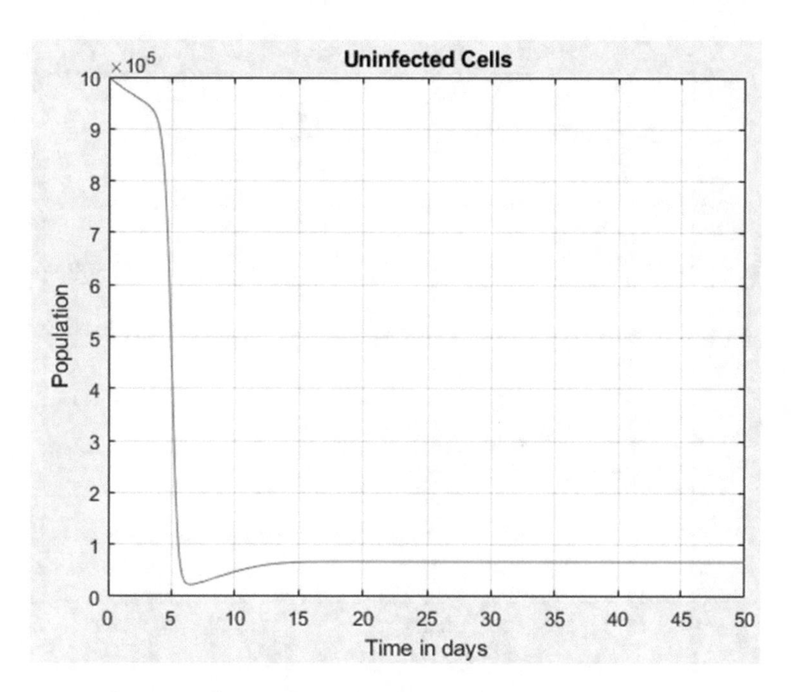

Figure 8-19. *Population of uninfected cells*

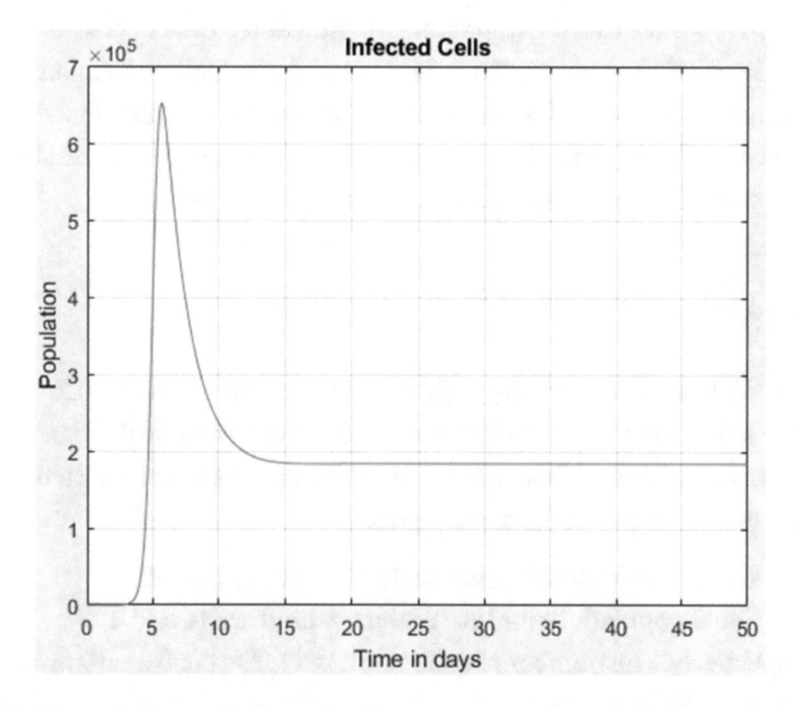

Figure 8-20. *Population of infected cells*

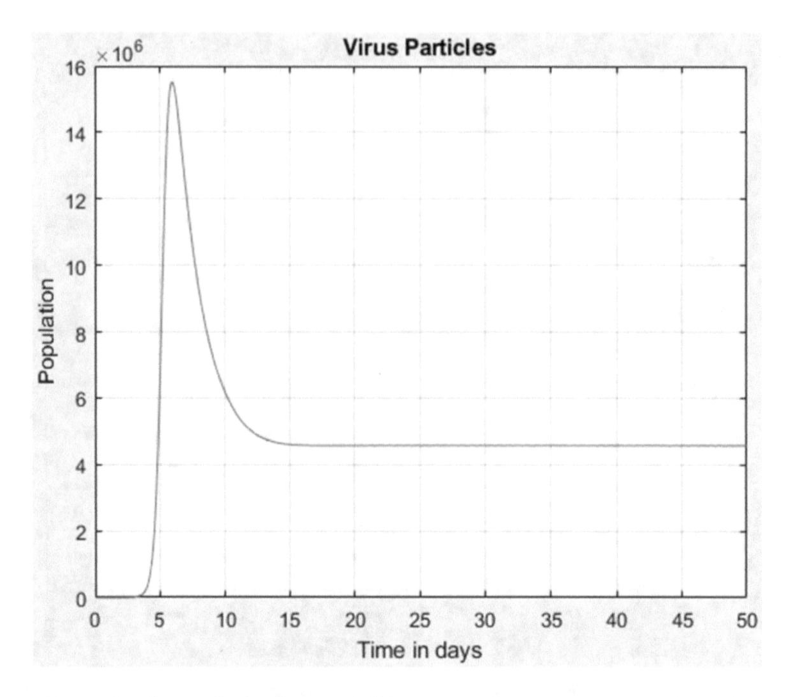

Figure 8-21. *Population of viral particles*

In Figure 8-19, we see that the population of uninfected cells decreased exponentially from 10^6 to about 67,000. After about 15 days, the population size becomes stable. For Figures 8-20 and 8-21, the scenarios look similar. Populations of both infected cells and virus particles exponentially increased from the third day to the sixth day. Then, after 15 days, their population sizes become stable as well.

Problems

8.1. Write code to solve the logistic equation with $K = 50$, $r = 0.2$, $N_0 = 4$, and time from 1 to 200. The code should animate the analytical solution. Then, both numerical and analytical solutions should be plotted on the same figure.

8.2. Write code to simulate the Nicholson-Bailey model. The initial populations for host and parasitoid are 10 and 5, respectively. The number of generation is 50. The coefficients are $e = 2$, $a = 0.0069$, and $r = 3$.

8.3. A DNA sequence is given as TTCCTACTTTTGTATTCGCCAG. Write a code to transcribe it to RNA.

8.4. A DNA sequence is given as GGTCGTAGCTAC. Write code to transcribe it to RNA. Then, by using the genetic code table, translate it to a sequence of amino acids. Finally, the code should print out the final sequence on the screen.

8.5. For the two sequences `Seq1=GTTAACAATT` and `Seq2=GTGAT`, write a code to print out the dot matrix and draw a dot plot. In the graph, matches should be displayed in red and mismatches should be shown in blue.

8.6. For Equations 10 and 11, write code to simulate the population sizes of the SIS model. The initial populations are 20 and 2 for S and I states, respectively. The time span is from 0 to 80. The coefficients are $\gamma = 0.50$ and $\beta = 0.8$.

8.7. For Equations 13, 14, and 15, write code to simulate the population sizes for all S, I, and R stages with respect to time. The initial populations are 20, 1, and 0 for S, I, and R populations, respectively. The time span is from 0 to 50. Other coefficients are $\gamma = 1.3$ and $\beta = 0.7$.

8.8. For Equations 17, 18, 19, and 20, write code to simulate the population sizes for all populations with respect to time. The initial populations are 27, 2, 1, and 0 for S, E, I, and R groups, respectively. The time span is from 0 to 50. Other coefficients are $\beta = 0.7$, $\alpha = 0.3$, and $\gamma = 0.2$.

CHAPTER 9

Signal Processing

Signal processing is one of the subfields of electrical engineering that analyzes, modifies, and operates information from all sources of data in the universe. A *signal* in signal processing is an electrical or electromagnetic current that relocates data from one device to another. Everything in the cosmos, such as sounds, images, and videos, can be described as a signal in the form of a wave. Signals can be represented as functions of one or more independent variables mathematically.

Signal processing has a very wide field of applications, including brain-computer interfaces, voice recognition, motion-sensing games, three-dimensional television, wearable technologies, biometric security, and more. Full coverage of this subject is beyond the scope of this chapter. Here, we review some introductory topics and a few applications in this area with MATLAB, including signal types, currents, Fast Fourier Transform, and harmonic analysis.

Signal Types

Signals can be classified into a few categories such as energy and power signals, or deterministic or probabilistic signals. In this section, we take a look at continuous and discrete signals, analog and digital signals, periodic and nonperiodic signals, and even and odd signals.

Continuous and Discrete Signals

For a signal defined as a function, if for every value of an independent variable (e.g., t), there exists a corresponding value (e.g., y), then that function is called a *continuous time* (CT) signal. However, if a signal represented as a function is defined for some discrete values of independent variable t, then it is called a *discrete time* (DT) signal. DT might

I. Turk, *Practical MATLAB*, https://doi.org/10.1007/978-1-4842-5281-9_9

not have a function of time in some cases. As an example, in a grayscale image, there are two-dimensional discrete values in the matrix of the image that represent colors of the image.

In Figure 9-1, we see the graphic of $y(t)$ where

$$y(t) = \sin(t), -2\pi \leq t \leq 2\pi \tag{1}$$

with a continuous signal.

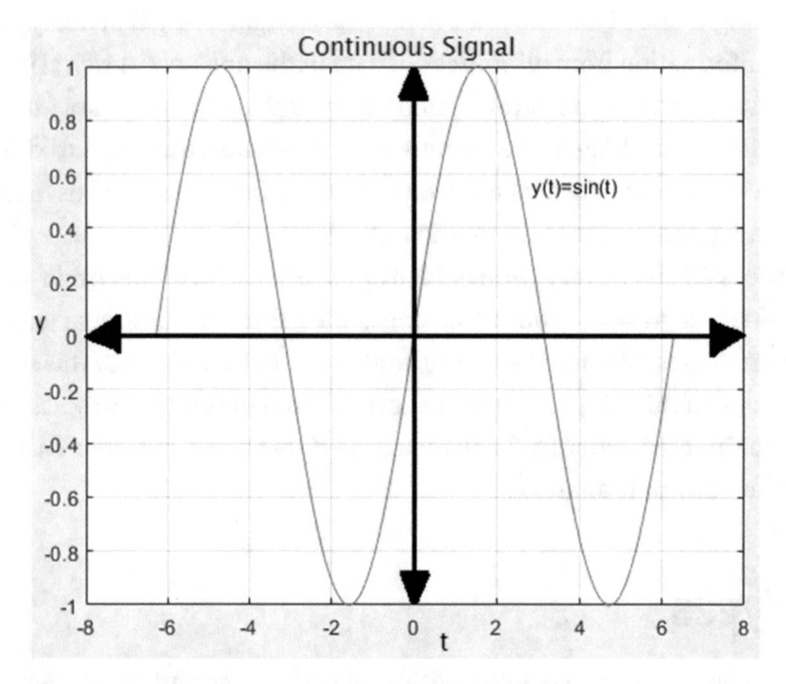

Figure 9-1. *An example of a continuous signal*

In Figure 9-2, although the same function is used, the output is different. Because discrete values of *t* are used, the obtained output is a discrete signal as well.

$$y[t] = \sin(t),\ t = 0,\ \pm\frac{\pi}{6},\ \pm\frac{2\pi}{6},\ \pm\frac{3\pi}{6},\ \ldots,\ \pm\frac{12\pi}{6} \tag{2}$$

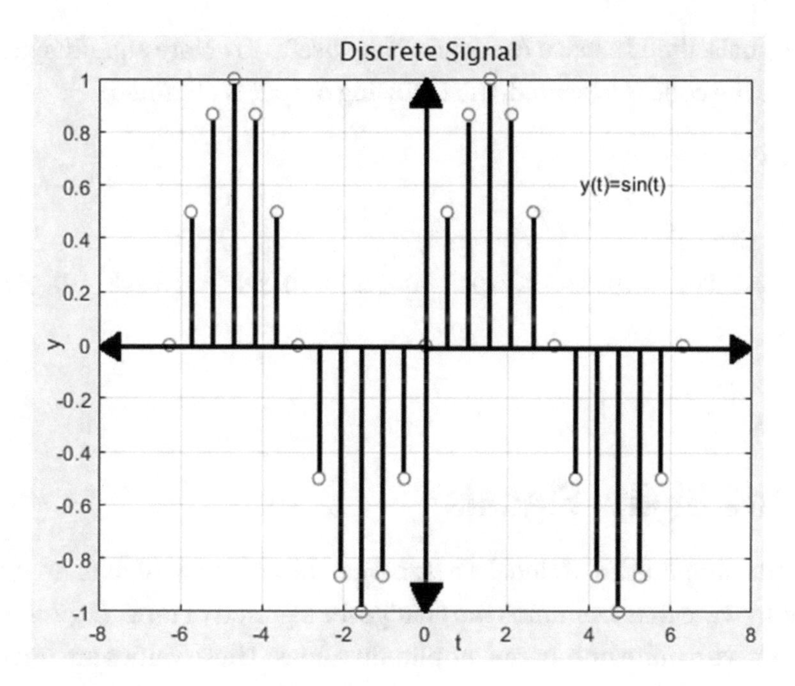

***Figure 9-2.** An example of a discrete signal*

In Equation 2, because t has discrete values, the function y has discrete values (corresponding to t values) as well.

In general, continuous functions are shown with parenthesis and square brackets are used to define discrete functions.

Example 9-1. For the given discrete signal function $y = f[t] = \cos(2t)$, where t is defined as $-\pi \leq t \leq \pi$, create 10 discrete equally spaced values for t. The code should then calculate the corresponding y values and print them on the screen.

Solution 9-1. The following code can be used to accomplish the given task.

Example9p1.m

```
%Example9p1
%This code has a discrete signal
t = linspace(-2*pi,2*pi,10);
y=cos(2*pi*t);
disp('y Values:');
disp(y);
```

For DC signals, the `linspace` function is very useful to create equally spaced values for time. Once the code is executed, the following output is obtained.

```
> Example9p1
y Values:
  Columns 1 through 9
   -0.2070  0.7580  -0.9983  0.8292  -0.3202  -0.3202  0.8292  -0.9983  0.7580
  Column 10
   -0.2070
>
```

Analog and Digital Signals

Considering the amplitudes of signals, a signal can be analog or digital. An *analog signal* is a continuous wave that maintains the change for a period of time. *Digital signals,* however, can have a finite number of amplitude values. These values are not continuous and infinite. In addition, the wave type of an analog signal looks like a sinusoidal function, whereas the wave of a digital signal looks like a square wave.

Some examples of analog signals are voices, music, temperature, and Wi-Fi signals in the air. Examples of digitals signal are voice recorded on a CD, a digital thermometer, and music in an MP3 player.

Periodic and Nonperiodic Signals

A signal that repeats itself after a period of time is called a *periodic signal.* A signal that does not repeat itself over a period of time is called an aperiodic or *nonperiodic* signal.

A continuous time signal $y(t)$ is called periodic if it satisfies the following equality.

$$y(t) = y(t + T_0) \tag{3}$$

for all t and positive constant T_0.

Similarly, a discrete time signal $y[t]$ is called periodic if it satisfies the following equality.

$$y[t] = y[t + T_0] \tag{4}$$

for all t and positive constant T_0.

Remark 9-1. Suppose that $y(t)$ is a function representing a continuous signal. Then if

$$y(t) = k * \sin^n(at + b), \tag{5}$$

or

$$y(t) = k * \cos^n(at + b), \tag{6}$$

then, the period , T_0, of signal $y(t)$ is

$$T_0 = \begin{cases} \dfrac{2\pi}{|a|}, \text{ if } n \text{ is an odd integer} \\ \dfrac{\pi}{|a|}, \text{ if } n \text{ is an even integer} \end{cases} \tag{7}$$

In Equations 5 and 6, constant values of k, a,and b are all real numbers, and n should be an integer. In Equation 7, $|a|$ is the absolute value of a.

Remark 9-2. Suppose that $y(t)$ is a function representing a continuous signal. Then if

$$y(t) = k * \tan^n(at + b), \tag{8}$$

or

$$y(t) = k * \cot^n(at + b), \tag{9}$$

then, the period T_0, of signal $y(t)$ is

$$T_0 = \frac{\pi}{|a|} \tag{10}$$

In Equations 8 and 9, constant values of k, a,and b are all real numbers, and n should be an integer.

Example 9-2. A CT signal is defined as $y(t) = 5sin^3(4t - 1)$. Find the period of the signal. Then write a code to plot the signal where $-2\pi \leq t \leq 2\pi$.

Solution 9-2. In the example, the signal is a continuous signal. Therefore, we can use Remark 9-1 to find the period. We see that $n = 3$, which is an odd number, and $a = 4$. Then $T_0 = \frac{2\pi}{|4|} = \frac{\pi}{2}$.

The following code can be used to accomplish the given task.

Example9p2.m

```
%Example9p2
%This code has a periodic signal
t = -2*pi:0.001:2*pi;
y=5*(sin(4*t-1)).^3;
plot(t,y);grid on;
xlabel('t');ylabel('y');
title('A Periodic Signal');
```

Once the code is run, the output shown in Figure 9-3 is the result.

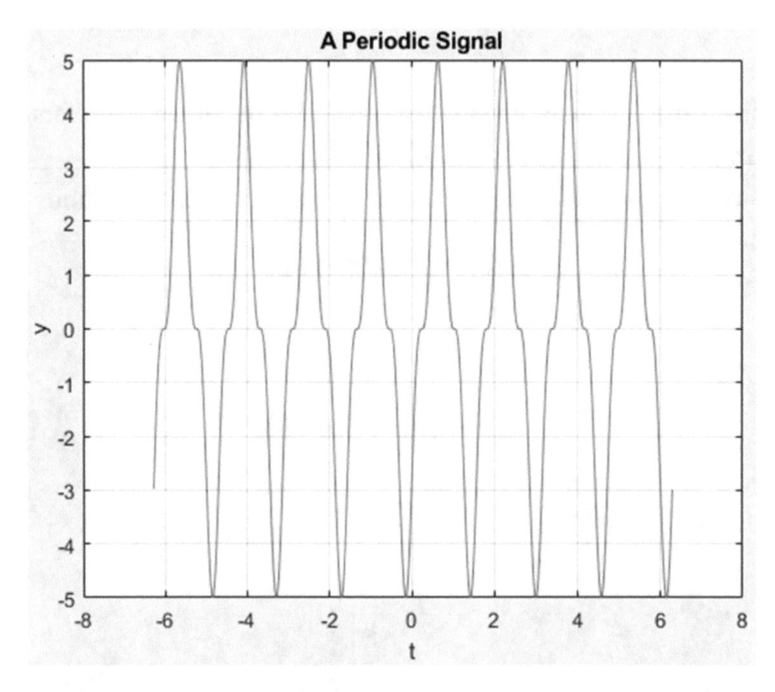

Figure 9-3. *Graph of $y(t) = 5sin^3(4t - 1)$*

As shown in Figure 9-3, the signal repeats itself after every $\frac{\pi}{2}$ value.

Example 9-3. A CT signal is defined as $y(t) = 2\,sin\,(4t) + \cos(6t - 2)$. Find the period of the signal. Then write a code to plot the signal where $-\pi \leq t \leq \pi$.

Solution 9-3. In the example, signal has a sine function added to the cosine. We will calculate the periods of these two functions separately. Then we will find the least common multiple (LCM) of those two periods. Let us define the period of sine term as T_0, and period of cosine term as T_1. Then by considering Remark 9-1, we see that $T_0 = \frac{2\pi}{|4|} = \frac{\pi}{2}$, and $T_1 = \frac{2\pi}{|6|} = \frac{\pi}{3}$. Then $\text{LCM}\left(\frac{\pi}{2}, \frac{\pi}{3}\right) = \pi$. The graph should have a period of π for the signal.

The following code can be used to accomplish the given task.

Example9p3.m

```
%Example9p3
%This code has a periodic signal
t = -pi:0.001:pi;
y=2*sin(4*t)+cos(6*t-2);
plot(t,y);grid on;
xlabel('t');ylabel('y');
title('A Periodic Signal');
```

Once the code is executed, we see the output shown in Figure 9-4.

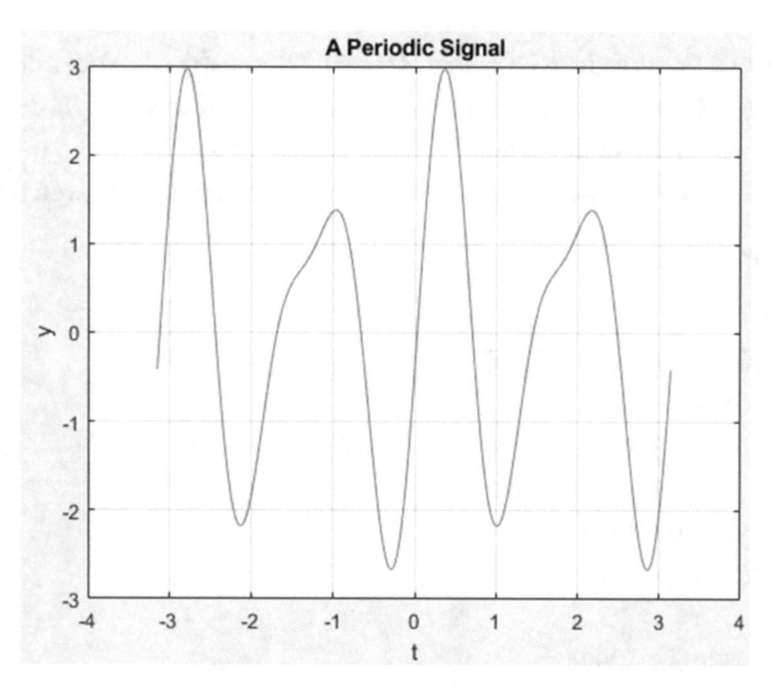

Figure 9-4. *Graph of $y(t) = 2sin(4t) + cos(6t - 2)$ signal*

Obviously, as shown in Figure 9-4, the signal repeats itself after every π period.

Even and Odd Signals

A continuous time signal $y(t)$ is called *even* if it satisfies the following equality.

$$y(-t) = y(t) \tag{11}$$

and it is called *odd* if it satisfies the following:

$$y(-t) = -y(t) \tag{12}$$

In a similar manner, a discrete time signal $y[t]$ is called even if it satisfies the following:

$$y[-t] = y[t] \tag{13}$$

and it is called odd if it satisfies the following:

$$y[-t] = -y[t] \tag{14}$$

Example 9-4. For the given CT signal $y(t) = sin^2(t) + \cos(t) + 1$, write a code to see whether the signal is even or odd over the graphic. Both graphics of $y(t)$, and $y(-t)$ should be shown on the same figure.

Solution 9-4. The following code can be used to accomplish the given task.

Example9p4.m

```
%Example9p4
%This code plots an even signal
t = -pi:0.001:pi;
y=(sin(t)).^2+cos(3*t)+1;
subplot(121);plot(t,y);grid on;
xlabel('t');ylabel('y');
title('Signal of y(t)');
t = -t;
y=(sin(t)).^2+cos(3*t)+1;
subplot(122);plot(t,y);grid on;
xlabel('t');ylabel('y');
title('Signal of y(-t)');
```

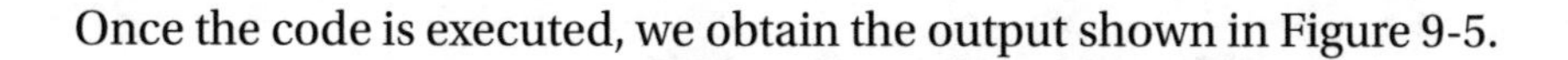

Once the code is executed, we obtain the output shown in Figure 9-5.

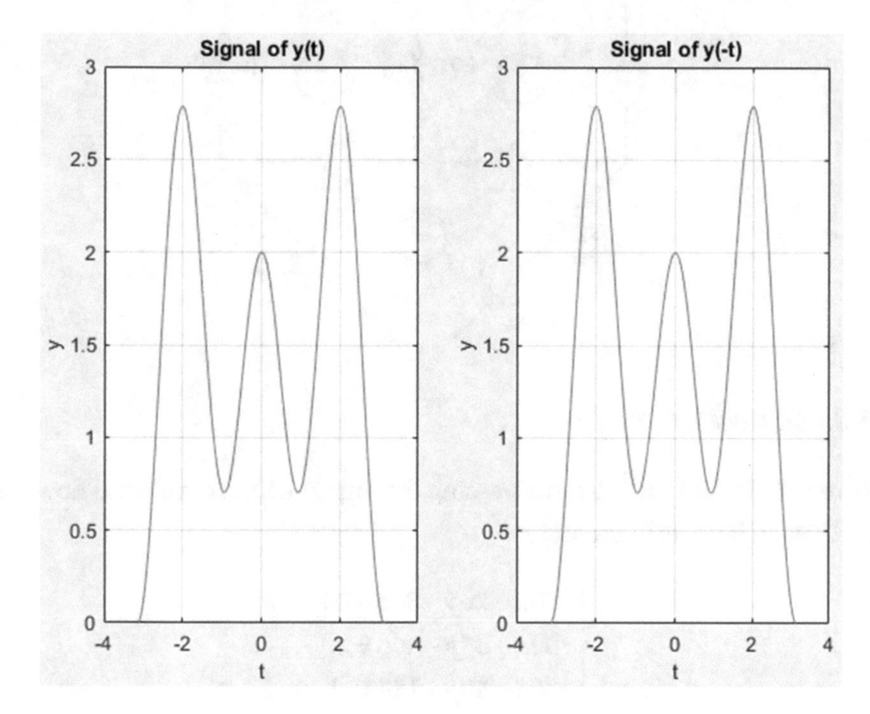

Figure 9-5. *Graph of $y(t) = sin^2(t) + cos(t) + 1$ signal*

Figure 9-5 shows that both drawings are exactly the same, which indicates that the signal is an even signal.

Electrical Currents

In this section, we learn how to find currents in an electrical network using an example. In addition, we examine how to find currents in an RL circuit considering Kirchhoff's laws.

Example 9-5. Consider an electrical network containing eight resistors and two batteries shown in Figure 9-6. Find the currents x, y, and z flowing through the loops.

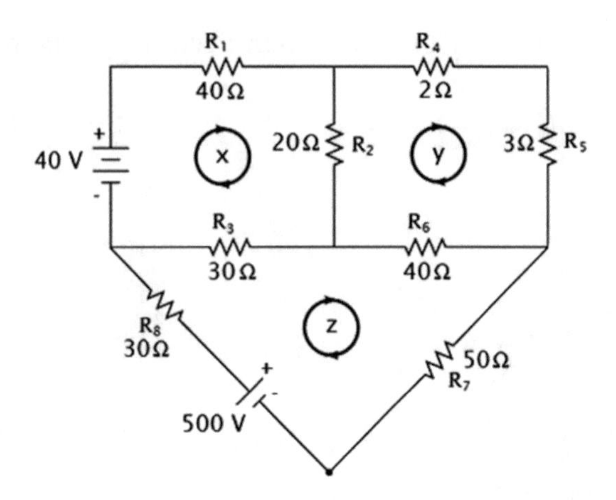

Figure 9-6. *An electrical network*

Solution 9-5. To find the currents passing through the loops, we write down the equations for each loop as following.

$$\begin{cases} 90x-20y-30z=40 \\ -20x+65y-40z=0 \\ -30x-40y+150z=500 \end{cases} \tag{15}$$

The following code can be used to solve this system of equations.

Example9p5.m

```
%Example9p5
%This code finds currents
%for an electrical network
A=[90 -20 -30;-20 65 -40;-30 -40 150];
B=[40;0;500];
Sol=(A)\B;
disp(['x=',num2str(Sol(1))]);
disp(['y=',num2str(Sol(2))]);
disp(['z=',num2str(Sol(3))]);
```

Once the code is executed, we obtain the following output.

```
> Example9p5
x=3
y=4
z=5
>
```

An electrical circuit can have a variety of components, including a switch, a voltage source (V), a resistor (R), and an inductor (L), as illustrated in Figure 9-7.

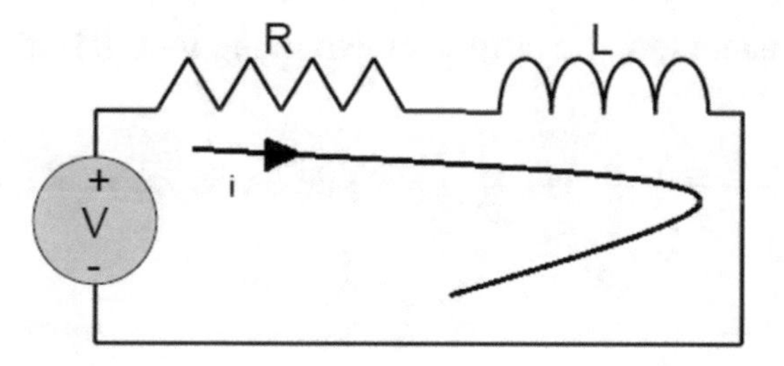

Figure 9-7. *An RL network*

The following is the governing equation of the network shown in Figure 9-7.

$$L\frac{di}{dt}+Ri=V \tag{16}$$

The solution of Equation 16 is

$$i=\frac{V}{R}\left(1-\exp\left(-R*\frac{t}{L}\right)\right) \tag{17}$$

Example 9-6. For an RL network, coefficients are given as $V=2$ Volt, $R=1$ Ohm; $L=1/2$ Henry, $0\le time\le 5$. Write code to simulate the values of current i with respect to time.

Solution 9-6: The following code can be used to accomplish the given task.

Example9p6.m

```
%Example9p6
%This code plots current values
t=0:0.05:5;
V=2;R=1;L=1/2;
i=(V/R)*(1-exp(-R.*t/L));
plot(t,i,'bo');grid on;
xlabel('t');ylabel('current');
title('Current with time');
```

Once the code is executed, the output shown in Figure 9-8 is displayed.

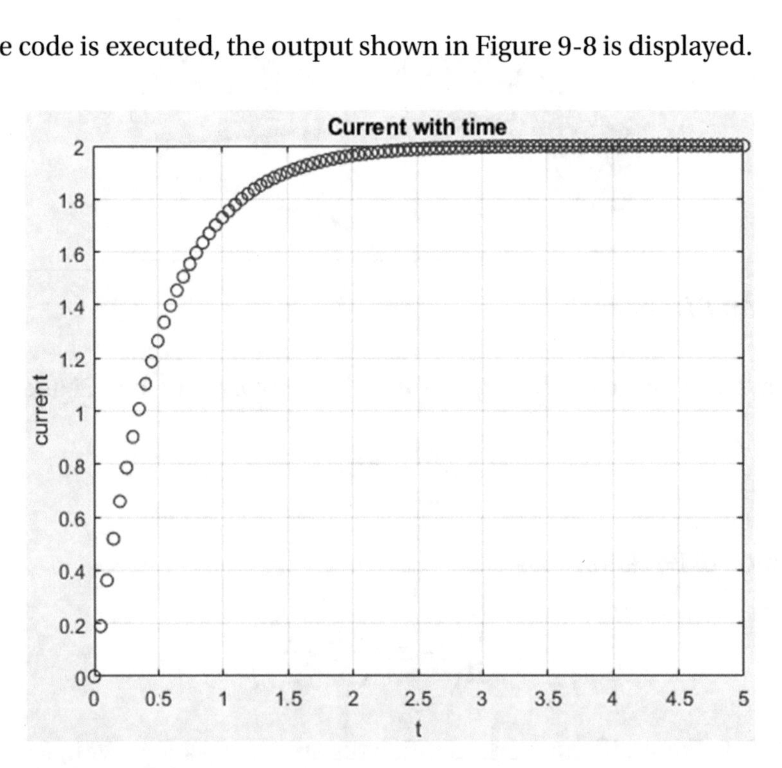

Figure 9-8. *Values of current in Example 9-6*

Harmonic Analysis

Periodic signals can be represented with a sum of sine and cosine waves using a Fourier series. Computing such signals is called *harmonic analysis.* At this point, we need to remember the definition of a Fourier series.

Remark 9-3. Suppose that for all x values, $f(x)$ is defined having period as 2L, meaning that $f(x) = f(x + 2L)$. Then the Fourier series of $f(x)$ is

$$f(x) \sim \frac{a_0}{2} + \sum_{n=1}^{\infty}\left(a_n \cos\left(\frac{n\pi x}{L}\right) + b_n \sin\left(\frac{n\pi x}{L}\right)\right). \tag{18}$$

Here, a_n and b_n are called the *Fourier coefficients* where

$$a_n = \frac{1}{L}\int_{-L}^{L} f(x)\cos\frac{n\pi x}{L}dx \tag{19}$$

and

$$b_n = \frac{1}{L}\int_{-L}^{L} f(x)\sin\frac{n\pi x}{L}dx \tag{20}$$

for $n = 1, 2, 3, \ldots$

Example 9-7. Coefficients of a signal $f(x)$ as defined in Equations 18, 19, and 20 are given as $L = 2$, $a_0 = 0$, $a_n = 0$, and $b_n = \begin{cases} \frac{4}{n\pi}, & n \text{ is odd} \\ 0, & n \text{ is even} \end{cases}$. Calculate the signal $f(x)$ for $n = 49$ as in Equation 18. Then plot the graphic of the signal $f(x)$ where $-2\pi \le x \le 2\pi$.

Solution 9-7. If we plug the given a_0, a_n, and b_n into Equation 18, we get the following:

$$f(x) = \frac{4}{\pi}\left(\sin\frac{\pi x}{2} + \frac{1}{3}\sin\frac{3\pi x}{2} + \frac{1}{5}\sin\frac{5\pi x}{2} + \frac{1}{7}\sin\frac{7\pi x}{2} + \ldots\right) \tag{21}$$

Therefore, we need to calculate Equation 21 for $f(x)$ having 49 terms in it. The following code can be used to get the graph of $f(x)$.

Example9p7.m

```
%Example9p7
%This code plots graph of a signal
n=49;Point=1000;
x=linspace(-2*pi,2*pi,Point);
bn=zeros(1,Point);a0=0*bn;an=0*bn;
```

```
for i=1:2:n
    Calc=(1/i).*sin((i*pi.*x)/2);
    bn = bn + Calc;
end
bn=(4/pi)*bn;fx=a0+an+bn;
plot(x,fx);grid on;axis('equal');
xlabel('x');ylabel('signal');
title('Square Wave Signal');
```

Once the code is executed, the output is shown in Figure 9-9.

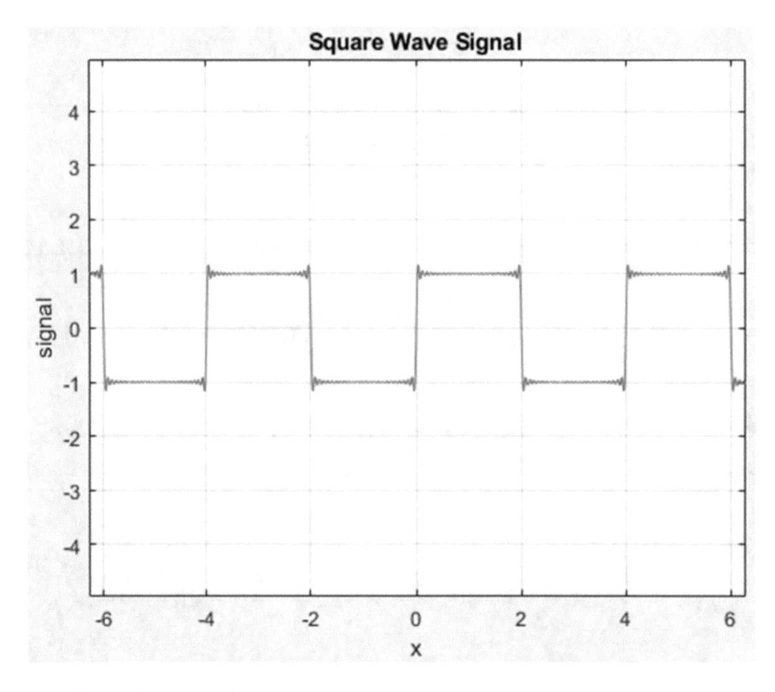

Figure 9-9. *Output of Example 9-7*

As shown in Figure 9-9, the signal is the square wave signal. The period of the signal is 4 ($L = 2$). If n is picked higher in the code, the image will look like exactly a square wave.

Example 9-8. Coefficients of a signal $f(x)$ as defined in Equations 18, 19, and 20 are given as $L = \frac{3}{2}$, $a_0 = 0$, $a_n = b_n = \begin{cases} \frac{4}{n\pi}, & n \text{ is odd} \\ 0, & n \text{ is even} \end{cases}$. Calculate the signal $f(x)$ for $n = 79$ as in Equation 18. Then plot the graphic of the signal $f(x)$ where $-2\pi \le x \le 2\pi$.

Solution 9-8. If we plug the given a_0, a_n, and b_n into Equation 18, we get the following:

$$f(x)=\frac{4}{\pi}\left(\cos\frac{\pi x}{1.5}+\sin\frac{\pi x}{1.5}+\frac{1}{3}\cos\frac{3\pi x}{1.5}+\frac{1}{3}\sin\frac{3\pi x}{1.5}+\ldots\right) \tag{22}$$

Therefore, we need to calculate Equation 22 for $f(x)$ having 79 terms in it. The following code can be used to generate the graph of $f(x)$.

Example9p8.m

```
%Example9p8
%This code plots graph of a signal
n=79;Point=1000;
x=linspace(-2*pi,2*pi,Point);
bn=zeros(1,Point);a0=0*bn;an=0*bn;
for i=1:2:n
    Calc1=(1/i).*cos((i*pi.*x)/1.5);
    Calc2=(1/i).*sin((i*pi.*x)/1.5);
    an = an + Calc1;
    bn = bn + Calc2;
end
fx=a0+(4/pi)*(an+bn);
plot(x,fx);grid on;axis('equal');
xlabel('x');ylabel('signal');
title('Signal in time');
```

Once the code is executed, we obtain the output shown in Figure 9-10.

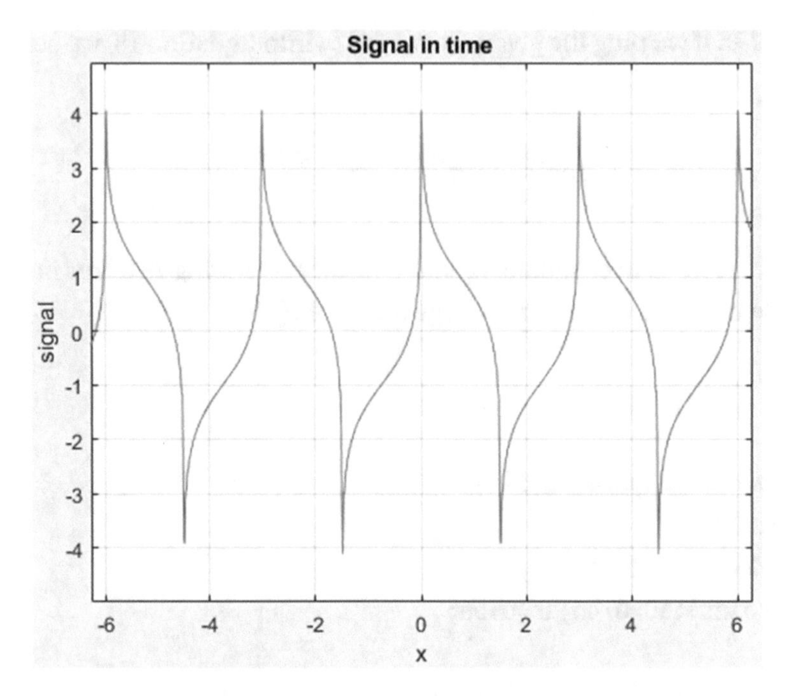

Figure 9-10. *Output of Example 9-8*

As shown in Figure 9-10, the signal has a period of 3 ($L = 1.5$).

Fast Fourier Transform

We already learned about Fourier series in the previous section on periodic functions. When it comes to work with nonperiodic signals, we need to know the Fourier transform method. A Fourier transform pairs a time series with the series of frequencies. Fast Fourier Transform (FFT) is a better way to handle discrete Fourier transforms (DFTs). Before understanding FFT, though, we will take a look at the definition of a Fourier transform.

Remark 9-4. The continuous time Fourier transform of a function $f(t)$ is given by:

$$F(k) = \int_{-\infty}^{\infty} f(t)e^{-2\pi ikt}dt, k \in R \tag{23}$$

$$f(t) = \int_{-\infty}^{\infty} F(k)e^{2\pi ikt}dk \tag{24}$$

where $F(k)$ can be obtained by using an inverse Fourier transform. The signal $f(t)$ has a Fourier transform if the result of Equation 23 converges. An inverse Fourier transform maps the series of frequencies back into the relevant time series.

Remark 9-5. A DFT of a time-limited sequence $f[t]$ where $0 \le t \le N-1$ is given by:

$$\text{Forward DFT } F[r] = \sum_{n=0}^{N-1} f[t] e^{\frac{-2\pi jnr}{N}}, 0 \le r \le N-1 \tag{25}$$

$$\text{Inverse DFT } f[t] = \frac{1}{N} \sum_{n=0}^{N-1} F[r] e^{\frac{-2\pi jnr}{N}}, 0 \le k \le N-1 \tag{26}$$

Although Equations 23 and 24 look very complex, by just substituting Euler's method where $e^{ix} = \cos(x) + isin(x)$, we can make them much easier, similar to Fourier series.

FFT is an algorithm that finds a solution to the DFT of a sequence in a better way. The algorithm was codeveloped by James W. Cooley and John W. Tukey in 1965. Cooley and Tukey took advantage of periodic sinusoids. In other words, some pieces of the calculations repeat themselves. By saving these computational costs, they reduced the time required and obtained a better algorithm.

For a signal with length *n,* time complexity (the time required to process *n* samples) is $O(n^2)$ using DFT. This complexity is much better in FFT than DFT, which is $O(nlog(n))$.

In MATLAB, the `fft` function is used to calculate FFT. Because indexing starts with 1, not 0, boundaries in the summation symbol for both Equations 25 and 26 are from $n = 1$ to N.

Example 9-9. Write code that applies `fft` to function x where $x = \sin(2\pi t)$, $0 \le t \le 1$, The code should generate 100 samples for t, and the length of vector x should be 512. The code should plot the graph of x, and power spectrum on the screen.

Solution 9-9. The following code can be used to accomplish the given task.

Example9p9.m

```
%Example9p9
%This code works with fft
Fs = 100; % Sampling frequency
t=linspace(0,1,Fs);
x = sin(2*pi*t);
Nfft = 512; % Length of FFT
X = fft(x,Nfft);
% FFT is symmetric, take half
```

```
X = X(1:Nfft/2);
% Frequency vector
f = (0:Nfft/2-1)*Fs/Nfft;
figure(1);%for graph of sine
stem(t,x);grid on;
title('Signal of Sine Wave');
xlabel('Time');
ylabel('Amplitude');
figure(2); % for power spectum
plot(f,abs(X));grid on;
title('Power Spectrum');
ylabel('Power');
xlabel('Frequency');
```

Once the code is run, the output shown in Figure 9-11 is the result.

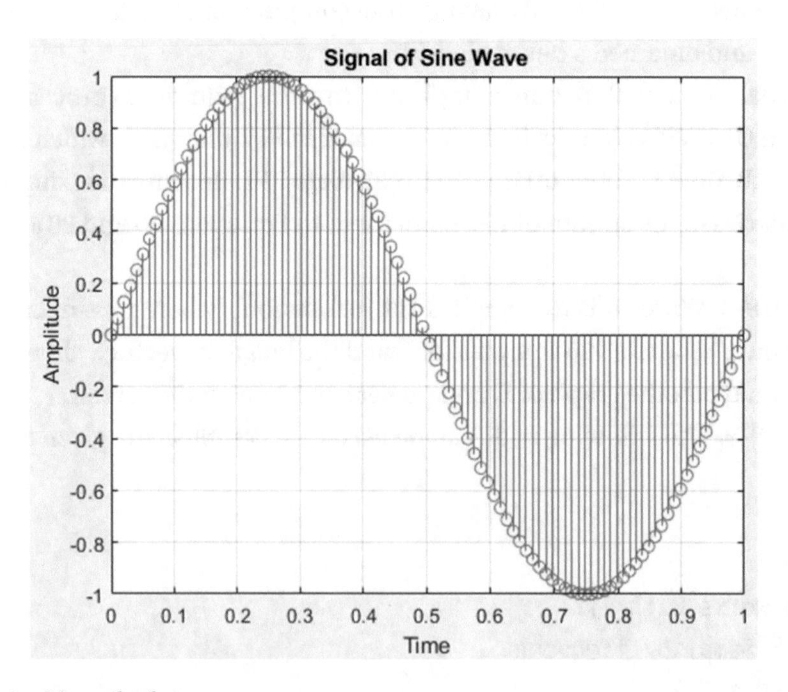

Figure 9-11. *Signal of sine wave*

For the power spectrum of the sine wave, we generated the output shown in Figure 9-12.

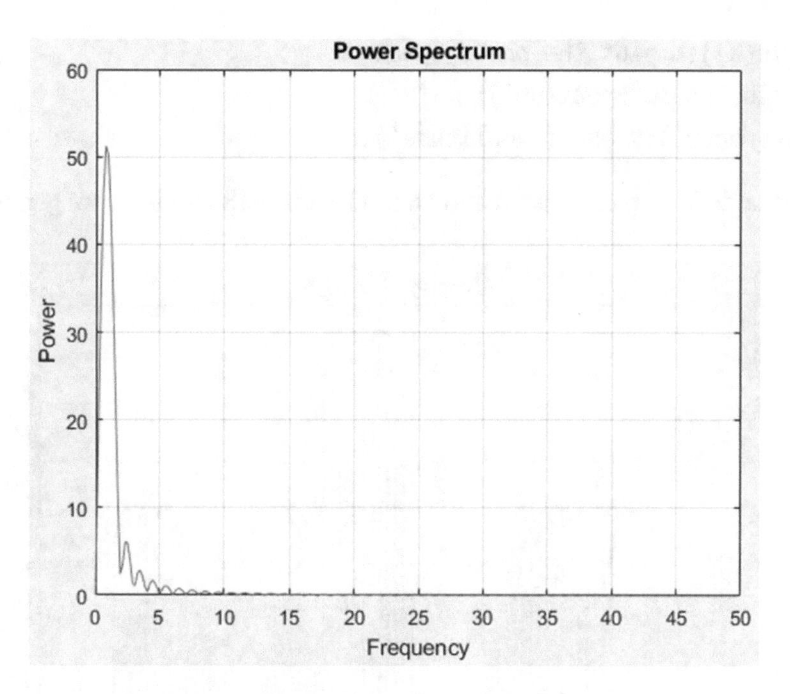

Figure 9-12. *Power spectrum of sine wave*

Example 9-10. Write code that applies `fft` to vector *x* where *x* has the first four elements as [1, 2, 0, 2] and 59 elements as zero. Then, the code should plot the estimation of the magnitude of the spectrum and the estimation of the phase spectrum on different figures.

Solution 9-10. The following code can be used to accomplish the given task.

Example9p10.m

```
%Example9p10
%This code works with fft
x = [1 2 0 2 zeros(1,60)]; % zero-padded sequence
N = length(x);X = fft(x);
X = fftshift(X); % shift DFT coefficients
w=linspace(-pi,pi,N);%compute frequencies
stem(w,abs(X)); % plot magnitude spectrum
grid on;title('Magnitude of Spectrum');
xlabel('Frequency');ylabel('Amplitude');
figure; % new one is opened
```

```
stem(w,angle(X));% plot the phase spectrum
grid on;title('Phase Spectrum');
xlabel('Frequency');ylabel('Amplitude');
```

Once the code is run, the output shown in Figures 9-13 and 9-14 are generated.

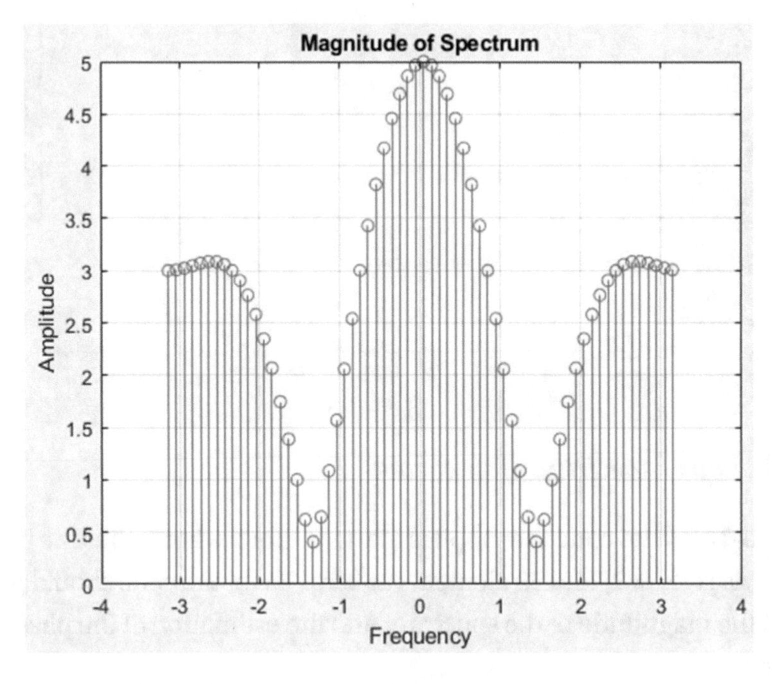

Figure 9-13. *Estimation of magnitude of spectrum*

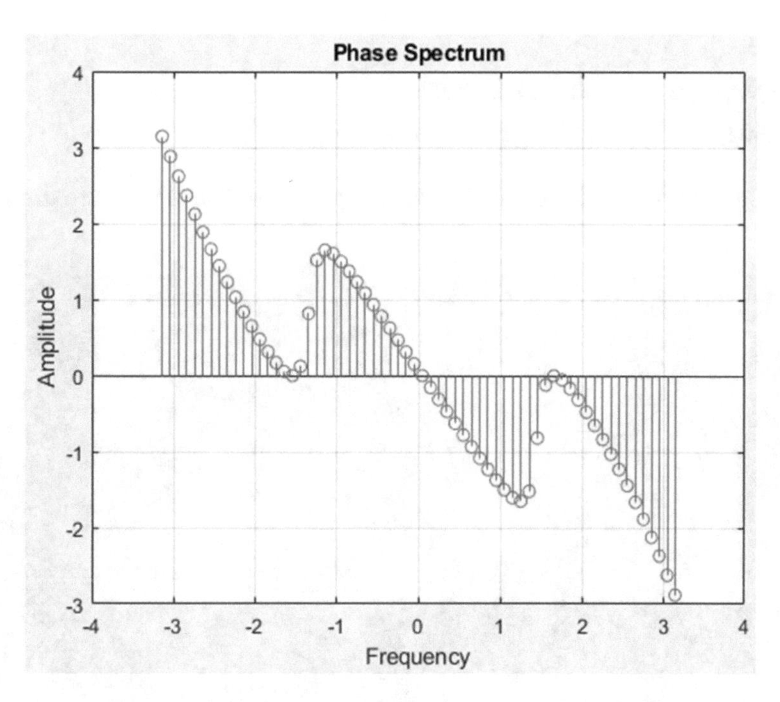

Figure 9-14. *Estimation of phase spectrum*

FFT can also be applied to audio files and image files for analysis. As an illustration, we show how to apply the `fft2` function to an image. Here, `fft2` is the two-dimensional function of `fft`. In this illustration, the `fft2` function is used as a filter, not for analysis. Working with image files is a broad topic that is examined in the next chapter.

Example 9-11. Write code that applies FFT to a colorful image as a filter. The name of the image is `covboys.png`, which is in the same directory as the codes. After applying FFT, the image should be shown on the screen. Then, the image should be converted back to the original one and the original image should be shown on a different figure as well.

Solution 9-11. The following code can be used to accomplish the given task.

Example9p11.m

```
%Example9p11
%This code apply fft to 2-dimensional picture.
P = imread('cowboys.png');%reads the image
Y=fft2(P); %apply the filter
imshow(Y) % filtered image is shown
```

```
figure;
YY=ifft2(Y);%image is converted back
imshow(YY/256);%original image is shown
```

Once the code is run, the output shown in Figures 9-15 and 9-16 are the result.

Figure 9-15. *Image after applying `fft2` function*

Figure 9-16. *After applying `ifft2` to get original image*

Problems

9.1. For the given discrete signal function $y = f[t] = \sin(2t)$, where t is defined as $-2\pi \le t \le 2\pi$, create 20 discrete equally spaced values for t. The code should calculate the corresponding y values and print them on the screen.

9.2. A CT signal is defined as $y(t) = 2sic(2t + 1)$. Find the period of the signal. Then write code to plot the signal where $-\pi \le t \le \pi$.

9.3. A CT signal is defined as $y(t) = sin(3t) + \cos(4t)$. Find the period of the signal. Then write a code to plot the signal where $-2\pi \le t \le 2\pi$.

9.4. For the given CT signal $y(t) = \cos(2t) + 10$, write code to see whether the signal is even or odd over the graphic. Both graphics of $y(t)$ and $y(-t)$ should be shown on the same figure.

9.5. Coefficients of a signal $f(x)$ as defined in Equations 18, 19, and 20 are given as $L = 3$, $a_0 = 0$, $a_n = 0$, and $b_n = \begin{cases} \dfrac{4}{n\pi}, & n \text{ is odd} \\ 0, & n \text{ is even} \end{cases}$.

Calculate the signal $f(x)$ for $n = 19$ as in Equation 18. Then plot the graphic of the signal $f(x)$ where $-\pi \le x \le \pi$.

9.6. Write a code that applies `fft` to function x where $x = \sin(2\pi t)$, $0 \le t \le 1$. The code should generate 200 samples for t and the length of vector x should be 1,024. The code should plot the graph of x and the power spectrum on the screen.

CHAPTER 10

Image Processing

MATLAB provides very powerful tools to work with graphics and image files, as well as for manipulating sounds, audio, and video files. This chapter deals with different topics related to image processing. At the beginning, I present the types of images. That is followed by discussions on converting image types and formats, operations on images, image enhancement, image restoration, color processing, image segmentation, and mathematical morphology.

Image Types

In MATLAB, most images are represented by two-dimensional matrices. In a matrix, every element corresponds to a pixel. If an image is composed of $n \times m$ pixels, its matrix will also have a size of $n \times m$. Hence, every single pixel in an image corresponds to an element in its matrix representation.

Let us think of an image that has two dimensions, $n \times m$. Its corresponding matrix will have n rows and m columns. If we call the matrix M, then its elements can be represented as shown in Figure 10-1.

$$
M(x,y) = \begin{bmatrix}
M(1,1) & M(1,2) & \ldots\ldots\ldots\ldots & M(1,m) \\
M(2,1) & M(2,2) & \ldots\ldots\ldots\ldots & M(2,m) \\
\ldots & \ldots & \ldots\ldots\ldots\ldots & \ldots\ldots \\
\ldots\ldots & \ldots\ldots & \ldots\ldots\ldots\ldots & \ldots\ldots \\
\ldots\ldots & \ldots\ldots & \ldots\ldots\ldots\ldots & \ldots\ldots \\
\ldots\ldots & \ldots\ldots & \ldots\ldots\ldots\ldots & \ldots\ldots \\
M(n,1) & M(n,2) & \ldots\ldots\ldots\ldots & M(n,m)
\end{bmatrix}
$$

Figure 10-1. *Matrix for a two-dimensional image*

I. Turk, *Practical MATLAB*, https://doi.org/10.1007/978-1-4842-5281-9_10

There is a remarkable number of supported image formats in MATLAB. These formats include `.bmp`, `.gif`, `.jpg`, `.jpeg`, `.hdf`, `.pcx`, `.png`, `.tiff`, `.xwd`, and more.

In the Image Processing Toolbox in MATLAB®, there are basically four different types of images: binary images, grayscale images, indexed images, and truecolor (RGB) images.

Binary Images

Binary images are composed of zeros and ones in their corresponding matrix representations. Zero represents the white color, and one represents the black color in the image. A binary image is stored as a logical array in the workspace. Figure 10-2 is an example of a binary image.

Figure 10-2. *An example of a binary image*

Grayscale Images

Grayscale images are represented by data matrices having various intensity values depending on the class of the matrix. Data types might belong to one of the classes `uint8`, `uint16`, `int16`, `single`, or `double`. Values for `single` or `double` data types range from 0 to 1. For `uint8`, the values range from 0 to 255. For `uint16`, the range is between 0 and 65,535. Finally, for `int16`, the values range from -32,768 to 32,767. Figure 10-3 is the grayscale version of the same image shown in Figure 10-2.

Figure 10-3. *An example of a grayscale image*

Indexed Images

An indexed image is composed of an array and a colormap matrix. The colormap matrix is an $m \times 3$ matrix of class `double`, where its elements range from 0 to 1. Each element in the array points to one of the rows in the colormap matrix to set the color of the images. The relationship between a colormap matrix and an array matrix is clearly shown in Figure 10-4.

Figure 10-4. *An indexed image with pixel values, index number, and colormap*

Truecolor (RGB) Images

Truecolor images are sometimes called RGB (Red-Green-Blue) images, as well. They are composed of three matrices, each one of which is of size $n \times m$. Accordingly, for a truecolor image (n, m, 1) shows the values for red, (n, m, 2) shows the values for green, and (n, m, 3) shows the values for blue. A combination of these three matrices determines the color of the truecolor (RGB) image. The array representing a truecolor image can be of class uint8, uint16, single, or double, where the pixel values specify the intensity values.

Obviously, RGB images represent the colors directly, rather than mapping the pixel values to a colormap matrix.

Example 10-1. Write a program that reads the pixel values of a truecolor image in the file named peppers.png. Then, take the transpose of red, green, and blue colors. Display the modified data as a new image.

Solution 10-1. The following code can be used to accomplish the given task.

Example10p1.m

```
%Example10p1
%This code takes the transpose of RGB
RGB = imread('peppers.png');
subplot(211);imshow(RGB)
New_RGB(:,:,1) = RGB(:,:,1)';
New_RGB(:,:,2) = RGB(:,:,2)';
New_RGB(:,:,3) = RGB(:,:,3)';
subplot(212);imshow(New_RGB)
```

This code reads the image peppers.png by using the imread command. Then the original image is shown in the first row of a figure. After that, transposes of each color—R, G, and B—are taken and assigned to a new variable named New_RGB. Finally, in the second row of the figure, the modified image is shown with the imshow function.

Once the code is executed, the image shown in Figure 10-5 appears.

Figure 10-5. *Output of Example10p1*

As you can see from Figure 10-5, the colors are preserved, but the direction of the image is altered.

Converting Image Types and Formats

Sometimes we might need to convert an image from one type to another. During this conversion, though, we might lose the quality of the original image.

There are several functions available for the conversion between class types, some of which are given in Table 10-1.

Table 10-1. *Useful Functions for Converting Class Types*

Function	Explanation
im2double	Converts image to double precision
im2int16	Converts image to 16-bit signed integers
im2single	Converts image to single precision
im2uint16	Converts image to 16-bit unsigned integers
im2uint8	Converts image to 8-bit unsigned integers

Table 10-2 provides a list of some of the useful functions for converting image types.

Table 10-2. *Useful Functions for Converting Image Types*

Function	Explanation
im2bw	Converts image to a binary image
ind2gray	Converts indexed image to grayscale image
gray2ind	Converts grayscale or binary image to indexed image
rgb2gray	Converts RGB image or colormap to grayscale
ind2rgb	Converts indexed image to RGB image

There are several functions for reading, writing, and performing other operations related to images, some of which are listed in Table 10-3.

Table 10-3. *Useful Functions Related to Images*

Function	Explanation
imread	Reads image from graphics file
imwrite	Writes image to graphics file
imfinfo	Gives information about graphics file
frame2im	Returns image data associated with movie frame
im2frame	Converts image to movie frame

Images can be saved in a different format using the `imwrite` function.

Example 10-2. Write a program that reads the image from the file `moon.tif`. Then, convert this image into `.jpg` format.

Solution 10-2. First, the following code reads the image. Then, the image is written in the target file format.

Example10p2.m

```
%Example10p2
%This code converts jpg to png format
Cowboy = imread('Cowboys2.jpg');
imwrite(Cowboy,'Cowboys2.png');
Cowboy_png = imread('Cowboys2.png');
imshow(Cowboy_png)
```

Once the code is executed, the image shown in Figure 10-6 is the result.

Figure 10-6. *Image converted from `.jpg` to `.png`*

Operations on Images

There are some functions that make users' job easier in manipulating images. Some of them are given in Table 10-4.

Table 10-4. *Useful Functions for Manipulating Images*

Function	Explanation	Example Format
imadd	Adds two images, or adds a constant to an image	`Z = imadd(X,Y)`
imrotate	Rotates the image A in degrees in counterclockwise direction	`B = imrotate(A,angle)`
imadjust	Adjusts the intensity values of the image	`J = imadjust(I)`
imresize	Resizes the image A to specified number of rows and columns	`B = imresize(A, [#rows #cols])`

Example 10-3. Write a program that reads the image from the file `football.jpg`. Then, rotate the image 30, 90, and 180 degrees.

Solution 10-3. The following code can be used to accomplish the given task.

Example10p3.m

```
%Example10p3
%This code rotates images
F = imread('football.jpg');
F30  = imrotate(F,30);
F90  = imrotate(F,90);
F180 = imrotate(F,180);
subplot(2,2,1),imshow(F)
title('Original Image')
subplot(2,2,2),imshow(F30)
title('Rotation with 30')
subplot(2,2,3),imshow(F90)
title('Rotation with 90')
subplot(2,2,4),imshow(F180)
title('Rotation with 180')
```

There are some images that already exist in MATLAB. One can directly read these images without having them in the directory. `Football.jpg` is one of these image files.

In the preceding code, the `imrotate` function rotates the images counterclockwise. Once the code is run, the output shown in Figure 10-7 is obtained.

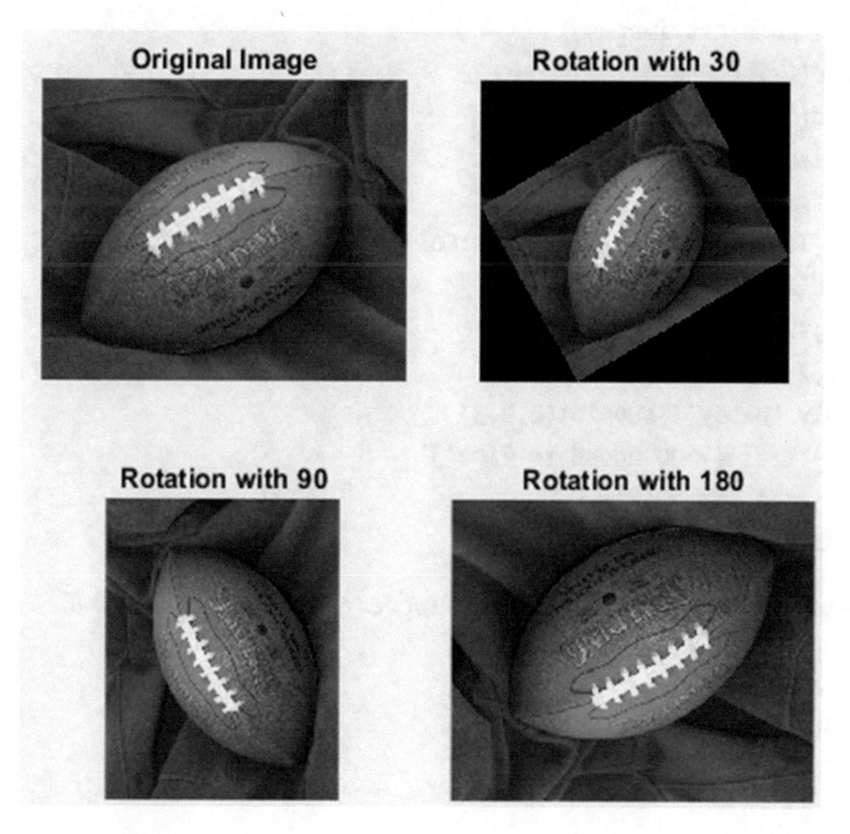

Figure 10-7. *Rotating images*

Example 10-4. Write a program that reads the images from the files `rice.png` and `cowboys.png`, respectively. Then, combine these images, and add 100 to the image `rice.png`. While combining images, dimensions of the pictures should match. If they do not match, then the code should pick the smallest row and column numbers from the dimensions of the pictures. The code should show all of these images on one figure separately.

Solution 10-4. Show the images `cowboys.png` and `rice.png` separately in a figure. Then, you can show the combined picture and the picture obtained by adding 100 to `rice.png` in the same figure. The code can be written as given shown here.

Example10p4.m

```
%Example10p4
%This code combines images
I = imread('rice.png');
J = imread('cowboys.png');
[rr,cr]=size(I);[rj,cj]=size(J);
Row=min(rr,rj);Col=min(cr,cj);
K = imadd(I(1:Row,1:Col),J(1:Row,1:Col));
JJ = imadd(I,100);subplot(2,2,1)
imshow(I);title('Only Rice')
subplot(2,2,2),imshow(J)
title('Only cowboy');subplot(2,2,3)
imshow(JJ);title('100 added to Rice')
subplot(2,2,4),imshow(K)
title('Combined Picture')
```

Once the code is executed, the images appear as shown in Figure 10-8.

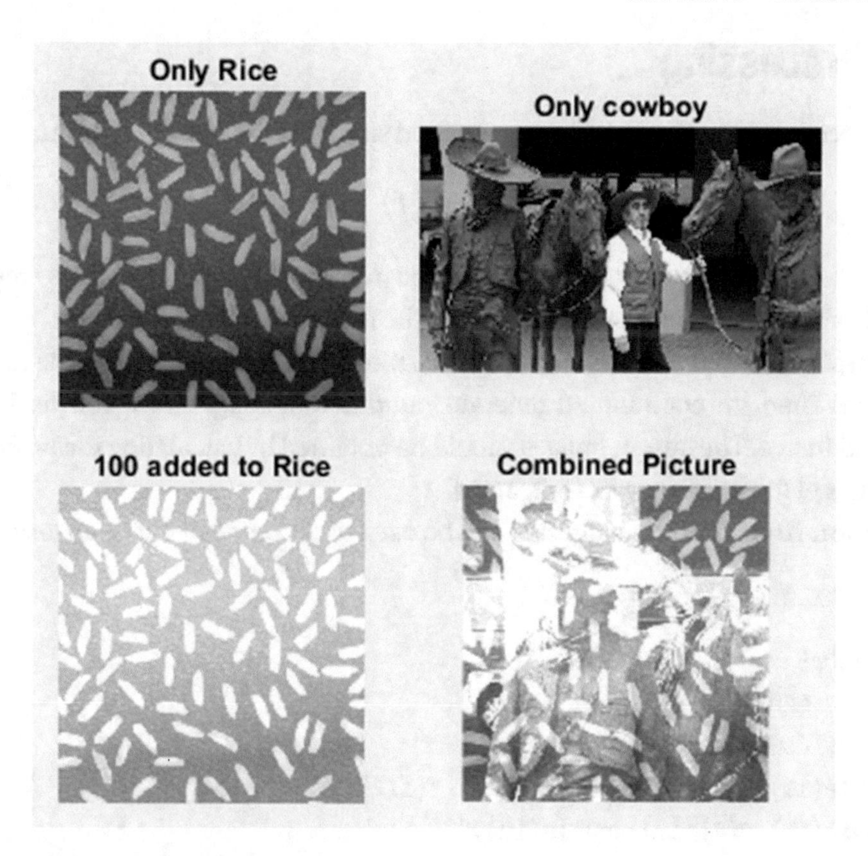

Figure 10-8. *Combined images*

As shown in Figure 10-8, after the addition, the combined images get brighter. This is due to the fact that having larger numbers in the matrix makes the image brighter.

Image Enhancement

The purpose of image enhancement is to modify the given image so that the revised image is improved. This can be done by manipulating the pixel value of the images, called *point processing*. The methods used can be from enhancement in the spatial domain or enhancement in the frequency domain.

Point Processing

Point processing operations can be represented simply in the following form:

$$g = T(f) \quad (1)$$

In Equation 1, g represents the output of the processed image pixel value, f represents the pixel value of the image, and T represents the point processing operation.

Example 10-5. Write a program that reads the image named `Cowboys.png` to the workspace. Then, the code should generate four different images using the pixel values of the read image. The output images should be obtained by calculating the powers of pixel values of the image as 1/5, ½, 2, and 4.

Solution 10-5. The following code can be used to accomplish the given task.

Example10p5.m

```
%Example10p5
%This code manipulates pixel values
im=imread('Cowboys.png');
im1=double(im).^(1/5);im2=double(im).^(1/2);
im3=double(im).^(2);im4=double(im).^(4);
im1=mat2gray(im1);im2=mat2gray(im2);
im3=mat2gray(im3);im4=mat2gray(im4);
subplot(221),imshow(im1);title('\^(1/5)');
subplot(222),imshow(im2);title('\^(1/2)');
subplot(223),imshow(im3);title('\^(2)');
subplot(224),imshow(im4);title('\^(4)');
```

In the code, it is not difficult to see that, if the power is smaller, then the obtained result becomes smaller, too. If the power increases, the result of the calculation increases, too. Before making calculations, pixel values of the image are converted to `double` with the `double` command. After calculating powers, they are sent back to intensity image values using the `mat2gray` function.

Once the code is run, the output shown in Figure 10-9 is displayed.

Figure 10-9. *Updated image with new pixel values*

Changing specific pixel values within an image is also possible, as illustrated in Example 10-6.

Example 10-6. Write a program that reads the image named `Cowboys.png` to the workspace, with an array matrix of 356 × 605 *x* 3. Then, for every element of the array having values between 30 and 150, change these values to 1.

Solution 10-6. The following code can be used to accomplish the given task.

Example10p6.m

```
%Example10p6
%This code changes pixel values in RGB
[RGB]=imread('Cowboys.png');
[Row,Col,ind]=size(RGB);
for i=1:ind
    for k=1:Row
        for m=1:Col
            if RGB(k,m,i)>=30 && RGB(k,m,i)<=150
```

```
                    RGB(k,m,i)=1;
                end
            end
        end
end
imshow(RGB)
```

Once the code is executed, the output is shown in Figure 10-10.

Figure 10-10. *Updated image with new pixel values*

The preceding code checks each cell value of R, G, and B cells. Then, 1 is assigned to elements having values between 30 and 150.

Using histograms help us to understand how the gray levels are distributed in a grayscale image. Therefore, important inferences can be drawn from the appearance of a histogram. To obtain the histogram of an image, the `imhist` function is used.

Example 10-7. In Example 10-5, there are four images plotted. Write code to show the histogram of those four images.

Solution 10-7. The following code can be used to accomplish the given task.

Example10p7.m

```
%Example10p7
%This code uses imhist function
im=imread('Cowboys.png');
im1=mat2gray(double(im).^(1/5));
im2=mat2gray(double(im).^(1/2));
im3=mat2gray(double(im).^(2));
im4=mat2gray(double(im).^(4));
subplot(221),imhist(im1);title('\^(1/5)');
subplot(222),imhist(im2);title('\^(1/2)');
subplot(223),imhist(im3);title('\^(2)');
subplot(224),imhist(im4);title('\^(4)');
```

Once the code is run, the histograms shown in Figure 10-11 are the result.

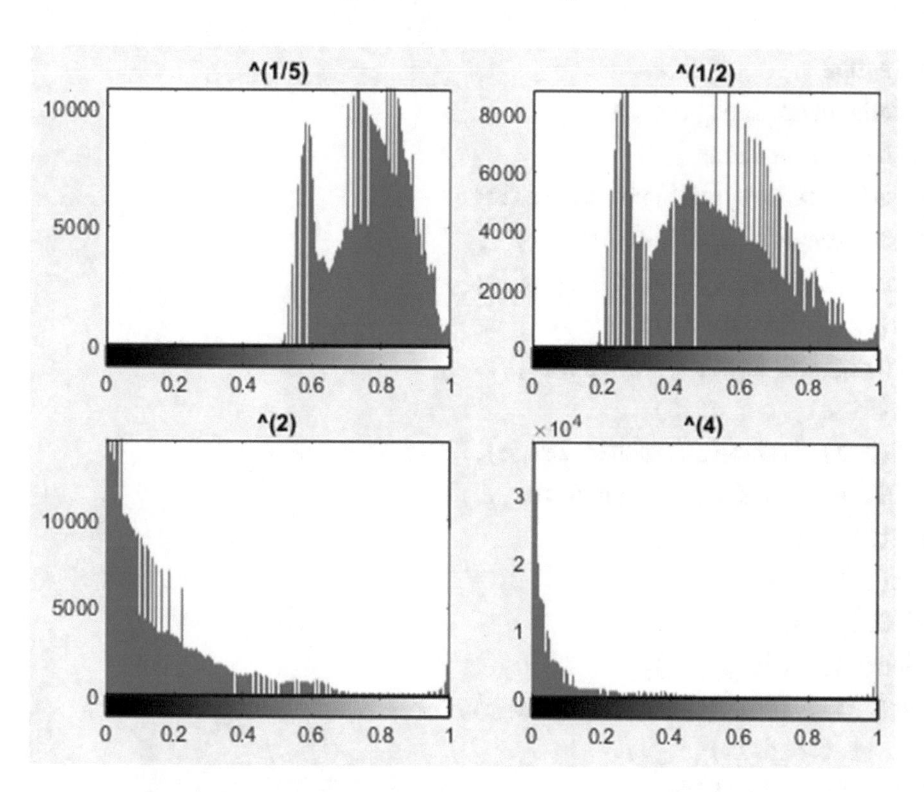

Figure 10-11. *Histograms of plots in Example 10-7*

As shown in Figure 10-11, lighter colors have values close to 1 and darker images have values close to zero. The number of bars used on the *x* axis is called *bin*. There are 256 bins by default when using the `imhist` function for grayscale images. If the image is binary, then 2 bins are used.

Contrast stretching is another technique used in image enhancement. To perform this technique, the `stretchlim` function is used together with the `imadjust` function. The `stretchlim` function computes a lower and upper limit of the image to use contrast stretching. Then, using the `imadjust` function, the intensity values of the image are adjusted.

Example 10-8. Write a code to adjust the pixel values of the image `Cowboys.png`. The code should saturate the upper 8% and lower 8% of pixel values. The original and adjusted image should be plotted on a figure, including their histograms.

Solution 10-8. The following code can be used to accomplish the given task.

Example10p8.m

```
%Example10p8
%This code uses imadjust and
%stretchlim functions
im=imread('Cowboys.png');fontSize=10;
Stretched_Image = imadjust(im, ...
    stretchlim(im, [0.08, 0.92]),[]);
subplot(221), imshow(im);
title('Original Image','FontSize',...
    fontSize);
subplot(222),imshow(Stretched_Image),
title('Stretched Image','FontSize',...
    fontSize);
subplot(223), imhist(im),
title('Original Image',...
    'FontSize',fontSize);
subplot(224), imhist(Stretched_Image),
title('Stretched Image',...
    'FontSize',fontSize);
```

Once the code is run, the result is shown in Figure 10-12.

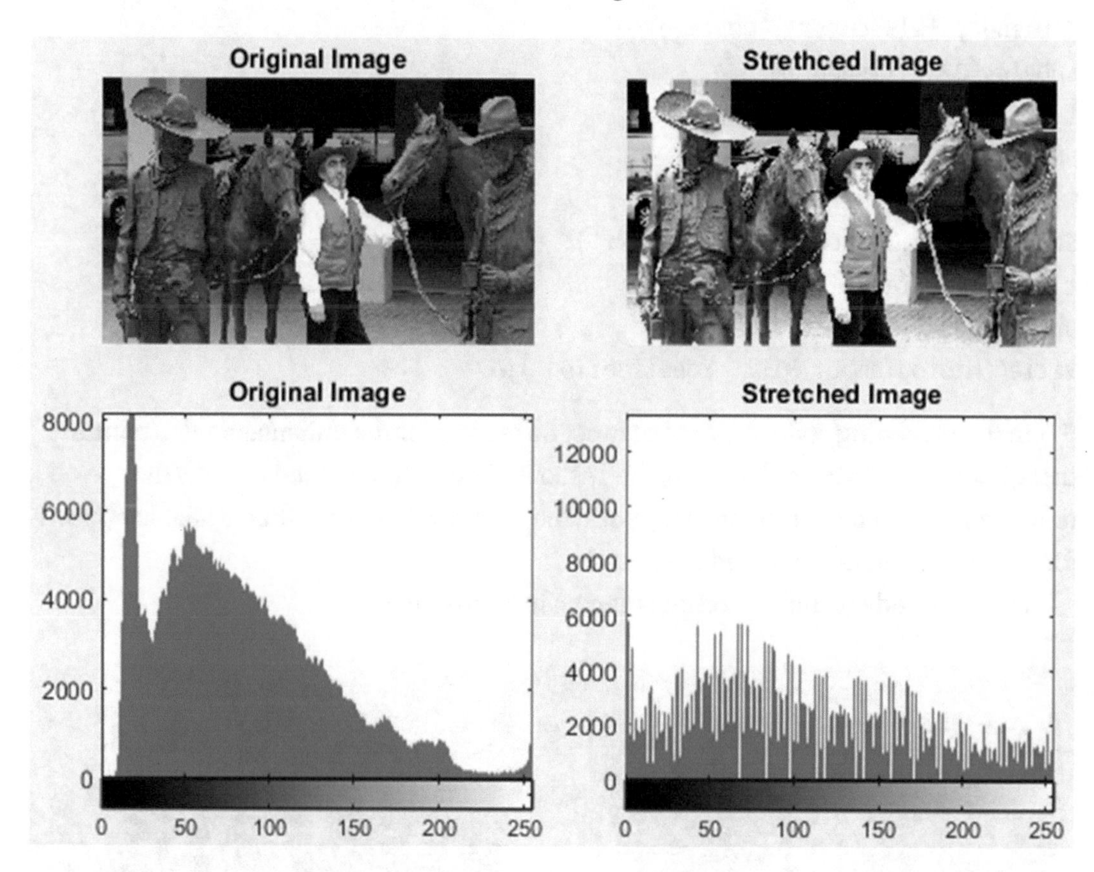

Figure 10-12. *Output of Example 10-8*

You can also enhance the contrast using histogram equalization.

Example 10-9. Write code to enhance the contrast of the `Cowboys.png` image using histogram equalization. The code should show the original image, enhanced image, the difference from the original to the enhanced image, and the histogram of the enhanced image on the same figure.

Solution 10-9. The following code can be used to accomplish the given task.

Example10p9.m

```
%Example10p9
%This code uses histeq and imshowpair func
im=imread('Cowboys.png');fontSize=10;
J = histeq(im);
imshowpair(im,J,'diff');
```

```
%Other Options of imshowpair-->'checkerboard'
%'blend','falsecolor','montage'
subplot(221),imshow(im)
title('Original Image');
subplot(222),imshow(J)
title('Histogram Equalization');
subplot(223),imshowpair(im,J,'diff')
title('Difference');
subplot(224),imhist(J)
title('Histogram of Hist. Equalization');
```

In the preceding code, the histeq function is used for the enhancement of contrast using histogram equalization. Also, the imshowpair function is used to show the difference between the original image and the enhanced image. Other available options for this command are included in the code.

Once the code is run, the result is shown in Figure 10-13.

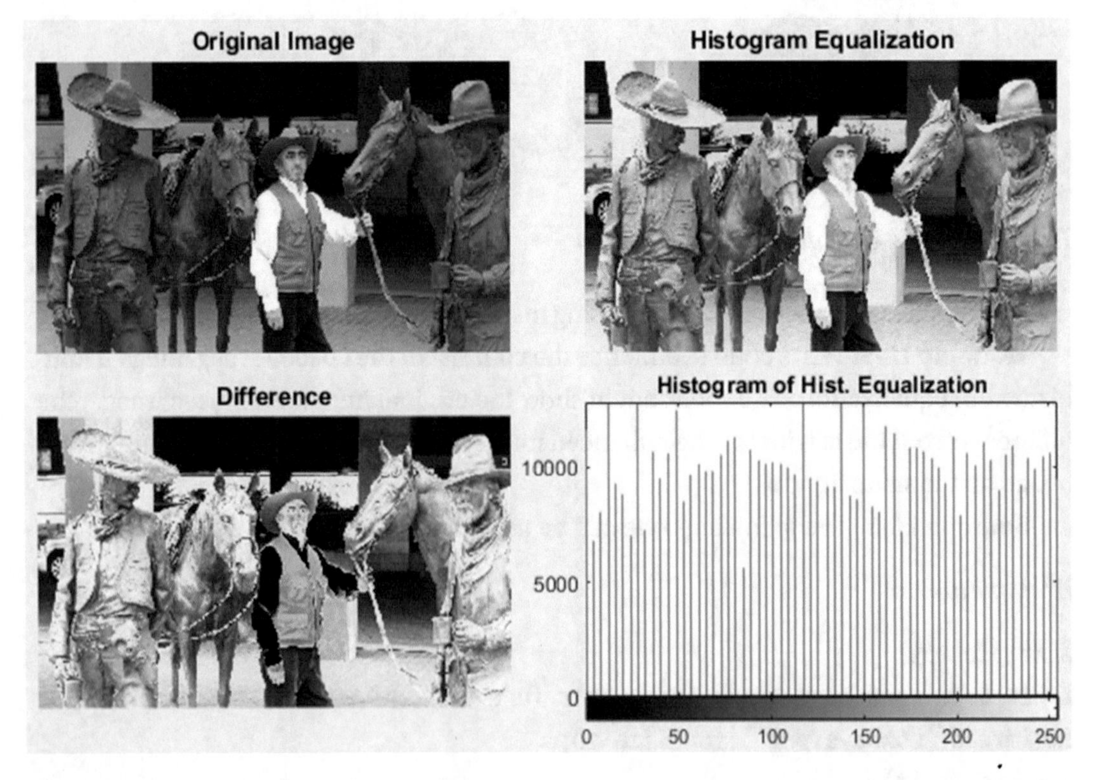

***Figure 10-13.** Output of Example 10-9*

Enhancement in the Spatial Domain

The spatial domain deals with the image plane itself. In this type of enhancement, manipulations are directly applied to the image pixels. The most common technique to enhance an image is filtering, a neighborhood operation that works on pixel values.

For a two-dimensional image, a filtering operation can be formulated as follows:

$$g(x,y)=T\left(f(x,y)\right) \tag{2}$$

where g represents the output of the processed values, f represents the pixel value of the image, and T represents the filter. If the operation applied on the pixels is linear, then the filter is said to be a linear spatial filter. Otherwise, the filter is said to be a nonlinear filter. Some commonly used filtering functions in the spatial domain are described in Table 10-5.

Table 10-5. *Some Basic Filtering Functions for the Spatial Domain*

Function	Explanation
imfilter	Filters the given array with given filter
fspecial	Creates 2-D filters for the types of `average`, `disk`, `gaussian`, `laplacian`, `log`, `motion`, `prewitt`, and `sobel`
imgaussfilt	Filters the given image with a 2-D Gaussian mask
medfilt2	Applies median filtering in two dimensions
medfilt3	Applies median filtering in three dimensions
ordfilt2	Applies order-statistic filtering in two dimensions
stdfilt	Applies standard deviation filtering
imboxfilt	Applies a 3-by-3 box filter to the image
conv2	Applies convolution to 2-D image

Example 10-10. Write a code to apply three different filters to an image `coins.png`. The filters are $[1,1,1;1,1,1;1,1,1]*1/9$, $[1,1,1,1,1;1,1,1,1,1;1,1,1,1,1;1,1,1,1,1;1,1,1,1,1]*1/25$, and $[0,1,0;1,-4,1;0,1,0]$. The filtered images and the original image should be shown in the same figure.

Solution 10-10. All given filters can be applied using the `imfilter` function. The following code can be used to accomplish the given task.

Example10p10.m

```
%Example10p10
%This code works with imfilter
H2=ones(3)/9;H3=ones(5)/25;
H4=[0, 1,0;1,-4,1;0, 1, 0];
I = imread('coins.png');
I2 = imfilter(I,H2);
I3 = imfilter(I,H3);
I4 = imfilter(I,H4);
subplot(2,2,1),imshow(I),title('Original Image');
subplot(222),imshow(I2),title('Filtering by H2');
subplot(223),imshow(I3),title('Filtering by H3');
subplot(224),imshow(I4),title('Filtering by H4');
```

Once the code is run, the resulting graphics is shown in Figure 10-14.

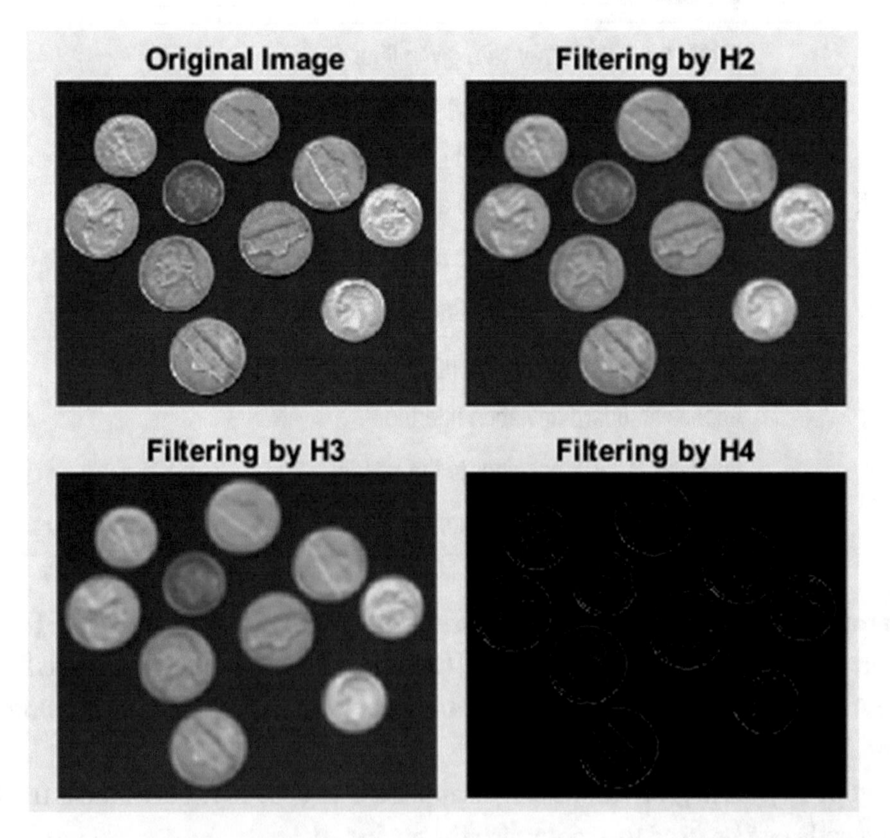

Figure 10-14. *Outputs of filters*

Example 10-11. Write a code to apply the filters of 'gaussian,' 'average,' and 'motion' to 'peppers.png' which MATLAB has. The filtered images and the original image should be shown on the same figure.

Solution 10-11. To apply the given special filters, they first need to be created using the `fspecial` function. Using the `imfilter` function, we can then apply them to the image. The following code can be used to accomplish the given task.

Example10p11.m

```
%Example10p11
%This code works with imfilter and fspecial
H2 = fspecial('gaussian');
H3 = fspecial('average');
H4 = fspecial('motion');
I = imread('peppers.png');
I2 = imfilter(I,H2);
I3 = imfilter(I,H3);
I4 = imfilter(I,H4);
subplot(221),imshow(I),title('Original Image');
subplot(222),imshow(I2),title('By gaussian');
subplot(223),imshow(I3),title('By average');
subplot(224),imshow(I4),title('By motion');
```

Once the code is run, the result shown in Figure 10-15 is obtained.

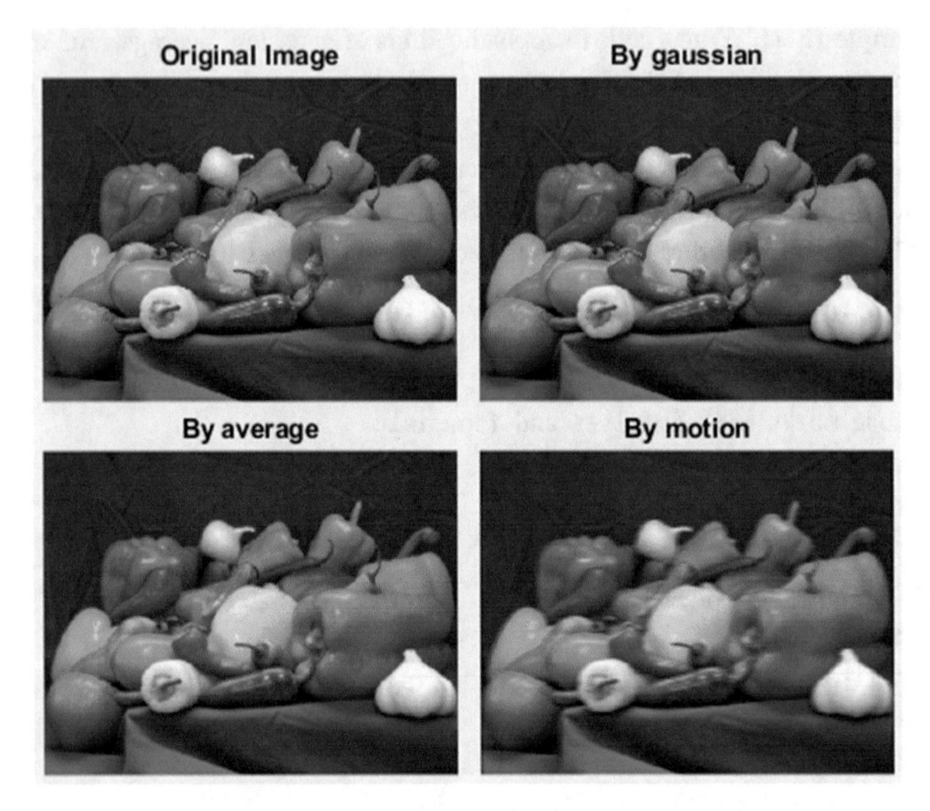

Figure 10-15. *Outputs of three special filters*

Enhancement in the Frequency Domain

The frequency domain deals with the rate of pixel change. These types of enhancements are based on modifying a Fourier transform. Some of the filtering functions used are listed in Table 10-6.

Table 10-6. *Filtering Functions for the Frequency Domain*

Function	Explanation
freqspace	Used to create frequency responses
freqz2	Creates frequency response for a 2-dimensional FIR filter
fsamp2	Creates a 2-dimensional FIR filter for an entered frequency response

The process of enhancement in the frequency domain can be summarized as follows.

1. Take the Fourier transform of an image.
2. Multiply the obtained output by a filter transfer function.
3. Take the inverse Fourier transform of the image

$$g(w_1,w_2)=F(w_1,w_2)*H(w_1,w_2) \tag{3}$$

These procedures are illustrated simply in Figure 10-16.

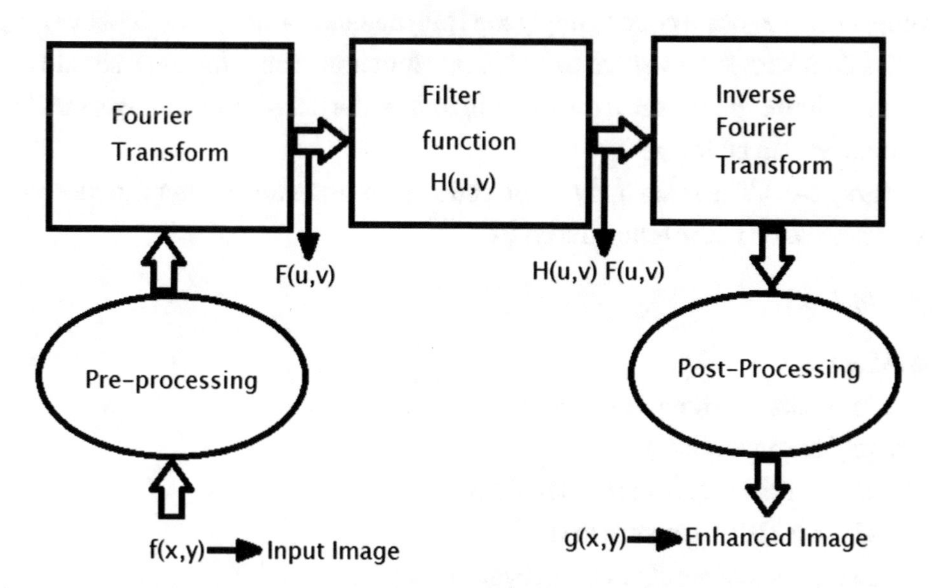

***Figure 10-16.** Diagram of filtering in the frequency domain*

Different filter functions can be selected as $H(u, v)$. For the low pass filter, an ideal low pass filter (ILPF) is used in general in two-dimensional applications.

$$H_{LP}(u,v)=\begin{cases}1, & if D(u,v)\leq D_0\\0, & if D(u,v)>D_0\end{cases} \tag{4}$$

In Equation 4, D_0 is the radius of the circle from the origin. The filter passes all the frequencies within D_0, and cuts off the rest of the frequencies. The simplest high pass filter is the complement of ILPF defined as follows.

$$H_{HP}(u,v)=1-H_{LP}(u,v) \tag{5}$$

The ideal high pass filter (IHPF) is used in general in two-dimensional applications.

$$H_{LP}(u,v)=\begin{cases}0, & if D(u,v)\leq D_0\\ 1, & if D(u,v)>D_0\end{cases} \tag{6}$$

Example 10-12. Write code to apply the ILPF defined earlier to the `Cowboys.png` image. The filter should be written as a separate function. The value of D_0 should be 50. The resulting figure should show the original image, the transformed image, the filter, and the final enhanced image.

Solution 10-12. The main body of the code and the separated filter function can be written as illustrated in the following code.

Example10p12.m

```
%Example10p12
%This code works in frequency domain
im =imread('cowboys.png');
P_grayscale = rgb2gray(im);close all;
P_grayscale=double(P_grayscale);
subplot(221),imshow(mat2gray(P_grayscale));
title('Original Image');
%image i frequency domain
freq_rec=fft2(P_grayscale);
freq_rec=fftshift(freq_rec);
freq_rec_disp=log(1+abs(freq_rec));
%image shown in frequency domain.
subplot(222),imshow(mat2gray(freq_rec_disp));
title('Fourier transform is applied');
[Row,Col]=size(P_grayscale);
Filter =My_filter(Row,Col);
% Applying filtering
```

```
Filtering = freq_rec.*Filter;
Filtering_disp=log(1+abs(Filtering));
subplot(223),imshow(mat2gray(Filtering_disp));
title('Applied Filter');
%%% converting back to image
FilteredIm=real(ifft2(ifftshift(Filtering)));
subplot(224),imshow(mat2gray(abs(FilteredIm)));
title('Enhanced Image');
```

My_filter.m

```
function Filter =My_filter(Roww,Coll)
Row=ceil(Roww/2);Col=ceil(Coll/2);
D_0=50;
Filter=zeros(Roww,Coll);
for m=1:Roww
    for n=1:Coll
        if((m-Row)^2+(n-Col)^2)^0.5<=D_0
            Filter(m,n)=255;
        end
    end
end%Filter = 255-Filter; can be tried
end
```

Once the code is run, the result is shown in Figure 10-17.

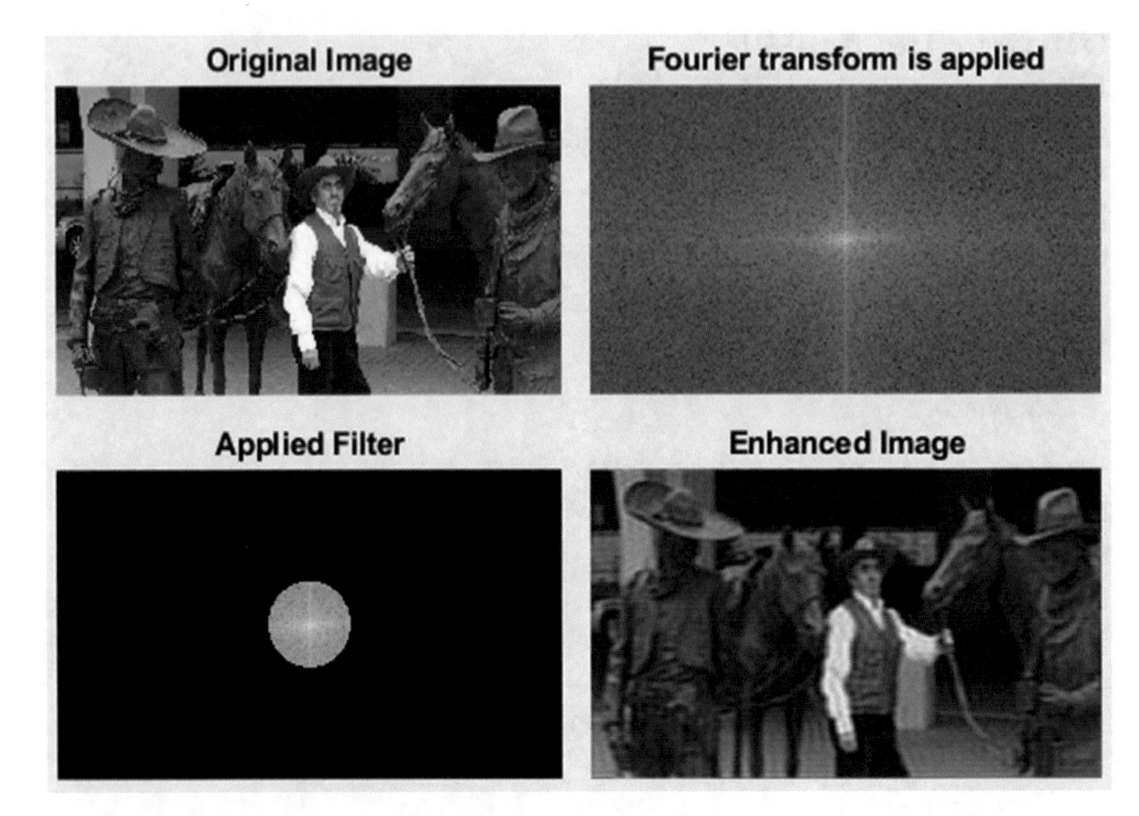

Figure 10-17. *Output of Example 10-12*

Enhancement with Other Functions

Enhancement can also be performed using integral image domain filtering, texture filtering, and edge preserving filtering functions, some of which are listed in Tables 10-7 and 10-8.

Table 10-7. *Filtering Functions Used in Integral Image Domain*

Function	Explanation
integralImage	Calculates integral image
integralImage3	Calculates 3-D integral image
integralBoxFilter	Filters the entered integral image with a 3-by-3 box filter

Table 10-8. *Filtering Functions Used for Edge Preserving*

Function	Explanation
imbilatfilt	Perfoms Gaussian bilateral filter on entered image
imguidedfilter	Applies filter using the guided filter
imnlmfilt	Performs a nonlocal means-based filter to color or grayscale images

Apart from the functions listed in Tables 10-7 and Table 10-8, `gabor` and `imgabor` functions are used in texture filtering.

Example 10-13. Write code that adds Gaussian white noise having zero mean and 0.0005 variance to the `peppers.png` image. Then use the `imnlmfilt` nonlocal means-based filter to remove the noise.

Solution 10-13. The following code can be used to accomplish the given task.

Example10p13.m

```
%Example10p13
%This code works with imnlmfilt function
I = rgb2gray(imread('peppers.png'));
noisyImage = imnoise(I,'gaussian',0,0.0005);
filteredImage = imnlmfilt(noisyImage);
subplot(211),imshow(noisyImage);
title('Image with noisy');
subplot(212),imshow(filteredImage);
title ('Filtered Image');
```

Once the code is run, the images shown in Figure 10-17 are obtained.

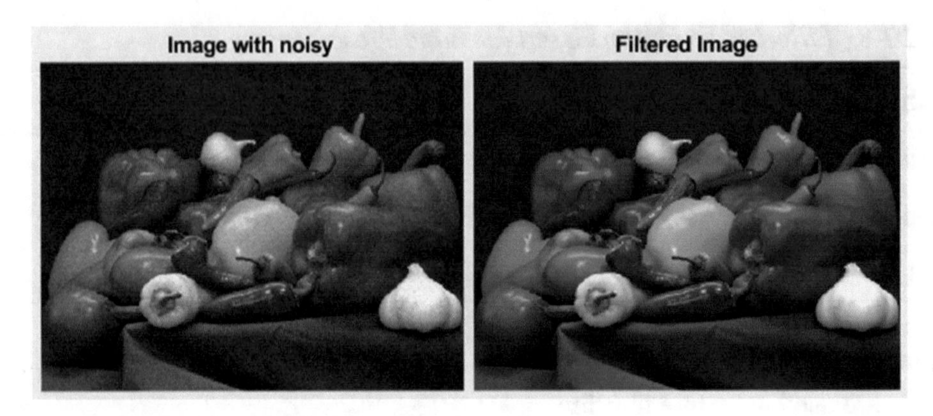

Figure 10-18. *Output of Example 10-13*

Image Restoration

Removing degradation from an image signal and regaining the original signal is called *image restoration.* In this section, we look at how the image signals lose their original state with salt and pepper noise, Gaussian noise, and periodic noise. Then methods of eliminating this noise are presented.

Adding and Removing Noise

Errors or noise can happen on images due to external factors such as transmission of image signals, or sending image signals via satellite.

Example 10-14. Write code that adds salt and pepper noise to the `peppers.png` image. Then use the `medfilt2` filter to remove the noise with 0.2 noise density.

Solution 10-14. The following code can be used to accomplish the given task.

Example10p14.m

```
%Example10p14
%This code applies salt & pepper noise
I1 = rgb2gray(imread('peppers.png'));
WithNois=imnoise(I1,'salt & pepper',0.2);
I2=medfilt2(I1);
subplot(121),imshow(WithNois);
title('Salt & Pepper Noise');
subplot(122),imshow(I2);
title ('Cleaning Noise');
```

Once the code is run, the images shown in Figure 10-19 are the result.

Figure 10-19. *Adding and cleaning salt and pepper noise*

Example 10-15. Write code that adds speckle multiplicative noise to the `peppers.png` image with noise density of 0.05. Then using the FIR filter and adaptive filter, remove the noise from the image. The code should generate the original, noisy, FIR filtered, and adaptive filtered images in the same figure.

Solution 10-15. The following code can be used to accomplish the given task.

Example10p15.m

```
%Example10p15
%This code applies and cleans noise
im = rgb2gray(imread('peppers.png'));
no1=imnoise(im,'speckle',0.05);
subplot(221),imshow(im),title('Original Image')
subplot(222),imshow(no1),title('Noisy Image')
no2=mat2gray(filter2(fspecial('average',3),no1));
subplot(223),imshow(no2)
title('FIR filtered Image')
no3= wiener2(no1,[5 5]);%neighborhoods of 5-by-5
subplot(224), imshow(no3)
title('Adaptive Filtered Image')
```

Once the code is run, the images shown in Figure 10-20 are generated.

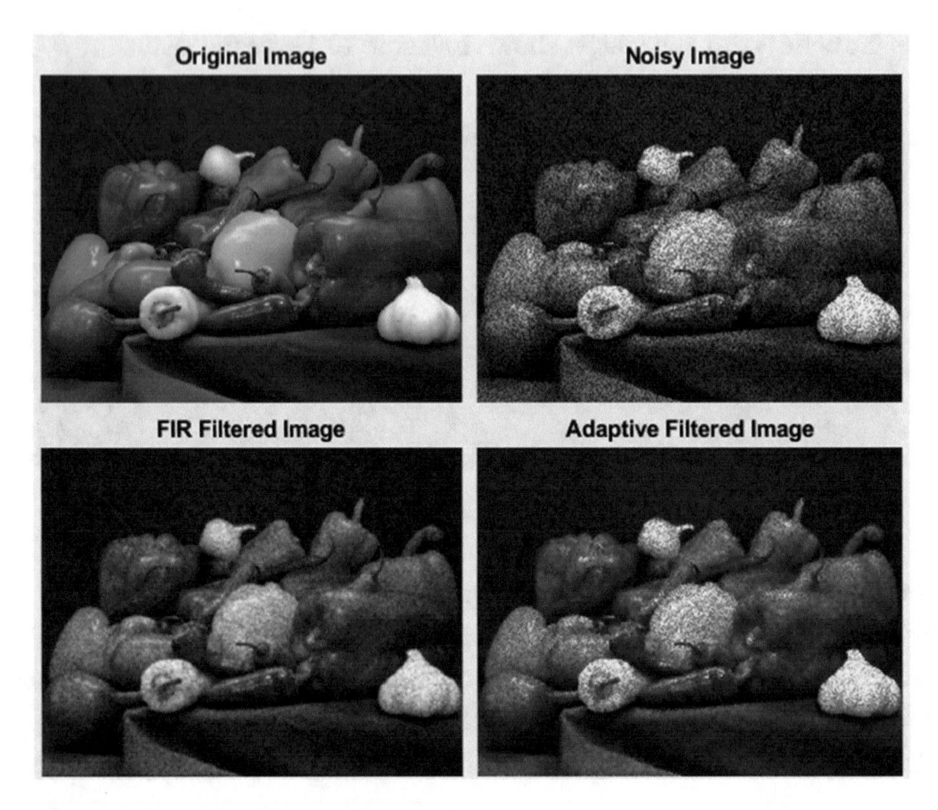

Figure 10-20. *Output of Example 10-15*

Color Processing

Color is a physical feature created by combined different frequencies of light when they are seen by the eyes. There are different definitions proposed by different scientists, but to keep things simple, I leave the definition short here.

In MATLAB, there are HSV, RGB, YIQ (defined by the National Television Systems Committee, or NTSC), YCbCr, and L∗a∗b color spaces. Primarily, though, operations are done in RGB space. For other spaces, there are conversion functions between RGB and other spaces such as `rgb2hsv`, `rgb2ntsc`, `rgb2ycbcr`, and `lab2rgb`. You can type `help` and one of these functions to get more information about that function in the command window of MATLAB.

In this section, we work with RGB colors.

Example 10-16. Write code that reads the `peppers.png` image and splits the RGB image into its red, green, and blue channels. The code should show the size of the image and the first five elements from each channel on the screen.

Solution 10-16. Using the `imsplit` function, the image can be separated into its red, green, and blue channels. The following code can be used to accomplish the given task.

Example10p16.m

```
%Example10p16
%This code separates RGB colors
RGB = imread('peppers.png');
[R,G,B]=imsplit(RGB);
disp(size(RGB));
disp(RGB(1:5,1:5,:));
subplot(221),imshow(RGB)
title('Original Image');
subplot(222),imshow(R),title('R Channel');
subplot(223),imshow(G),title('G Channel');
subplot(224),imshow(B),title('B Channel');
```

Once the code is run, the following output is obtained in the command window.

```
> Example10p16
   384   512     3
(:,:,1) =
   62   63   63   65   66
   63   61   59   64   63
   65   63   63   66   66
   63   67   67   63   64
   63   62   64   65   66
 (:,:,2) =
   29   31   34   30   27
   31   31   32   30   28
   29   30   31   30   31
   29   29   31   31   31
   31   32   33   30   31
```

```
(:,:,3) =
   64   64   64   60   59
   62   64   64   60   59
   60   62   63   61   61
   62   63   63   60   62
   62   63   62   61   61
>
```

As shown in the output, the size of image is 284 × 512 ×3, and the relevant array of the RGB image is a three-dimensional matrix. Each dimension corresponding to R, G, and B channels has a 284 × 512 matrix.

Figure 10-21 is the result. The original image is a combination of the other three images.

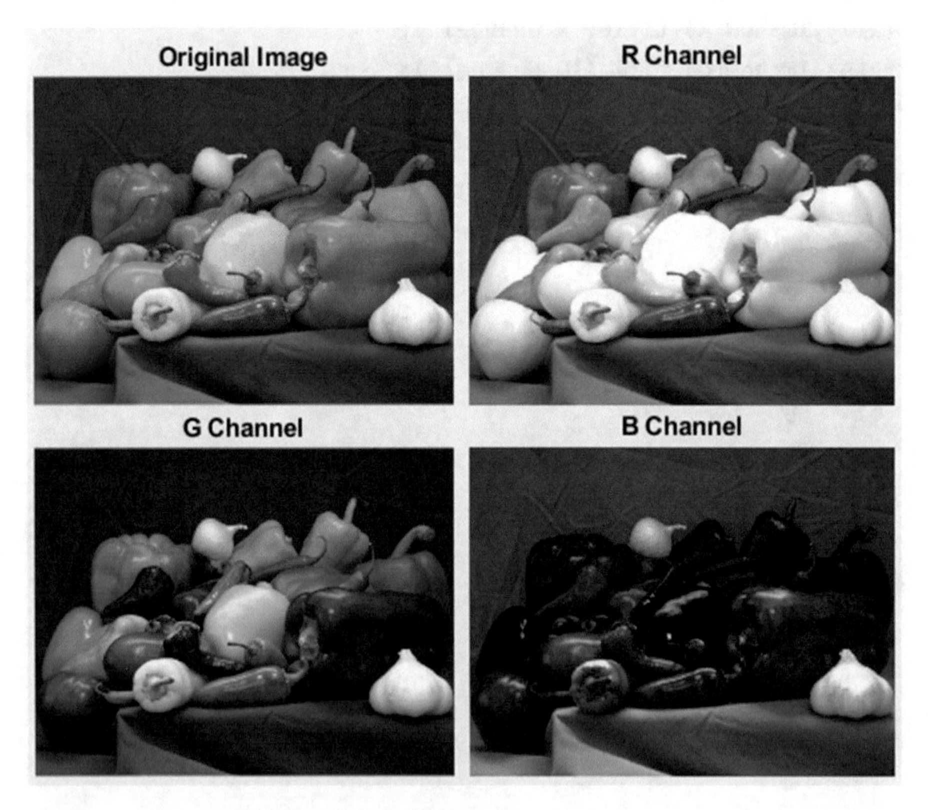

Figure 10-21. *Output of Example10-16*

An illustration of manipulating pixel values was presented in Example 10-6. Changing the coefficient of every channel in an RGB image is another option, as illustrated in Example 10-17.

Example 10-17. Write code that multiplies every channel with different coefficients of the `peppers.png` image. The first coefficients of the red, green, and blue channels are 1/3, 2/3, and 3/3; the second coefficients are 2/3, 3/3, and 1/3; and the last coefficients are 3/3, 1/3, and 2/3. After each channel is calculated with the given coefficients, channels should be combined to represent an RGB image. Therefore, there should be three color images obtained from different coefficients. The code should show three all of these images and the original image in one figure.

Solution 10-17. To keep the calculations for each channel, we can create a four-dimensional array. In this array, the first two dimensions belong to pixel values of each channel. The third dimension can keep the values for R, G, and B. The last dimension can keep the values with given coefficients. The following code can be used to accomplish the given task.

Example10p17.m

```
%Example10p17
%In code, channels different coefficients
RGB=imread('peppers.png');
New(:,:,1,1)=RGB(:,:,1)*(1/3);
New(:,:,2,1)=RGB(:,:,2)*(2/3);
New(:,:,3,1)=RGB(:,:,3)*(3/3);
New(:,:,1,2)=RGB(:,:,1)*(2/3);
New(:,:,2,2)=RGB(:,:,2)*(3/3);
New(:,:,3,2)=RGB(:,:,3)*(1/3);
New(:,:,1,3)=RGB(:,:,1)*(3/3);
New(:,:,2,3)=RGB(:,:,2)*(1/3);
New(:,:,3,3)=RGB(:,:,3)*(2/3);
subplot(221),imshow(RGB);
title('Original Image');
subplot(222),imshow(mat2gray(New(:,:,:,1)));
title('Image with R*(1/3)-G*(2/3)-B*1');
```

```
subplot(223),imshow(mat2gray(New(:,:,:,2)));
title('Image with R*(2/3)-G*(1)-B*(1/3)');
subplot(224),imshow(mat2gray(New(:,:,:,3)));
title('Image with R*(1)-G*(1/3)-B*(2/3)');
```

Once the code is run, the images shown in Figure 10-22 are obtained in the command window.

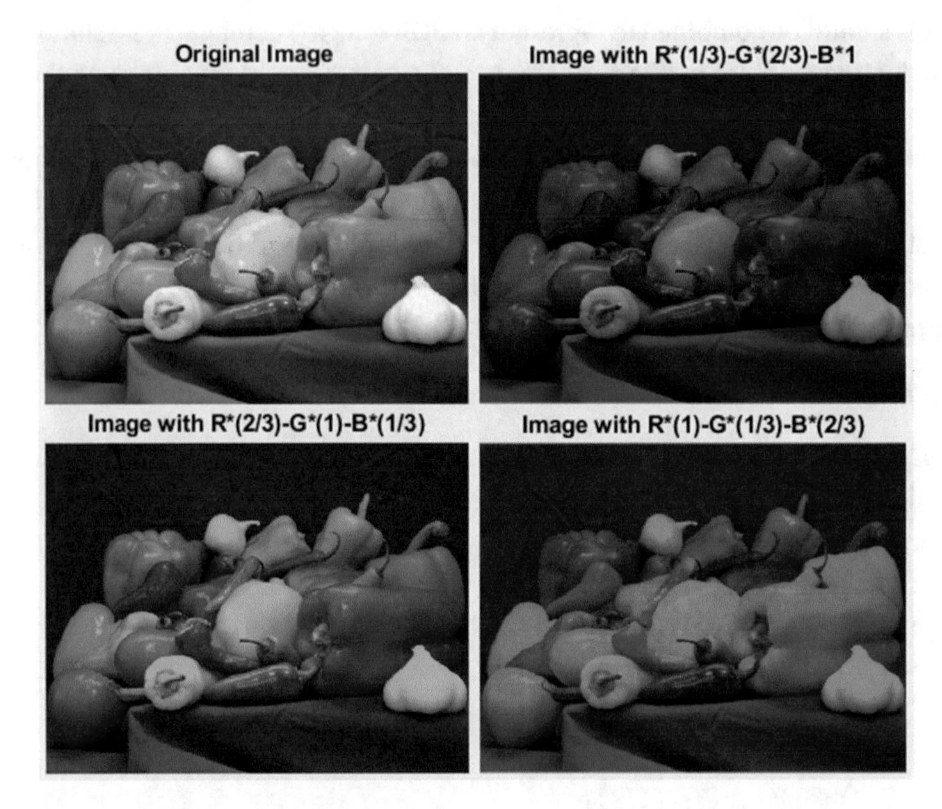

Figure 10-22. *Output of Example 10-17*

Image Segmentation

Image segmentation is the process of partitioning an image into constituent parts, or dividing an image into different regions. Generally, pixels that belong to the same segment are expected to have similar pixel values, and are thus expected to form a connected region. Common techniques for segmentation are thresholding, edge detection, and region-based methods.

Thresholding

Assigning zero or one to each pixel of a grayscale image considered an exact point or points is called thresholding. There are two types of thresholding: single and double.

Single Thresholding

In single thresholding, only one point is taken into account. Let us assume that the threshold value is defined as T. The pixel value at point (x,y) can be defined as follows:

$$\mathrm{P}(x,y)=\begin{cases}\text{white, if the value at }(x,y)>T\\ \text{black, if the value at }(x,y)\leq T\end{cases} \tag{7}$$

Therefore, in Equation 7, white or black coloring is defined for every pixel value of $\mathrm{P}(x, y)$ within the grayscale image.

Example 10-18. Write code to detect the borders of the `coin.png` file with $T = 0.61$ as the threshold.

Solution 10-18. The following code can be used to accomplish the given task.

Example10p18.m

```
%Example10p18
%This code uses thresholding
T=0.61;
im=mat2gray(imread('coins.png'));
[Row,Col]=size(im);
NewIm=zeros(Row,Col);
for i=1:Row
    for k=1:Col
        if im(i,k)>T
            NewIm(i,k)=1;
        else
            NewIm(i,k)=0;
        end
    end
end
figure,imshowpair(im,mat2gray(NewIm),'montage');
title ('Original Image and Thresholding');
```

Once the code is run, the output shown in Figure 10-23 is obtained.

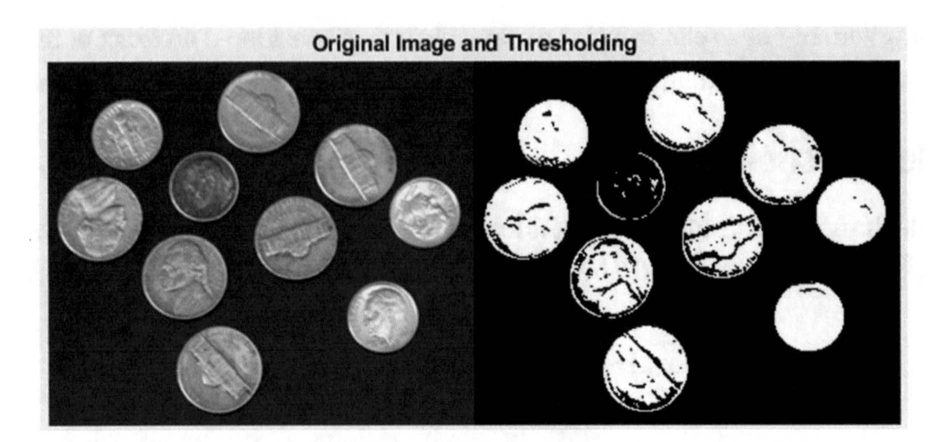

Figure 10-23. *Output of Example 10-18*

Similar thresholding can be performed using built-in MATLAB functions as well.

Example 10-19. Write code to detect the borders of the `coins.png` image using the `graythresh` function, which computes a global threshold using Otsu's method. To binarize the image, use the `imbinarize` built-in function.

Solution 10-19. The following code can be used to accomplish the given task.

Example10p19.m

```
%Example10p19
%This code uses graythresh and imbinarize functions
im = imread('coins.png');
Level = graythresh(im);%using Otsu's method
BinaryIm = imbinarize(im,Level);%binarize the image
imshowpair(im,BinaryIm,'montage');
title ('Original Image and Thresholding');
```

Once the code is run, the output shown in Figure 10-24 is the result.

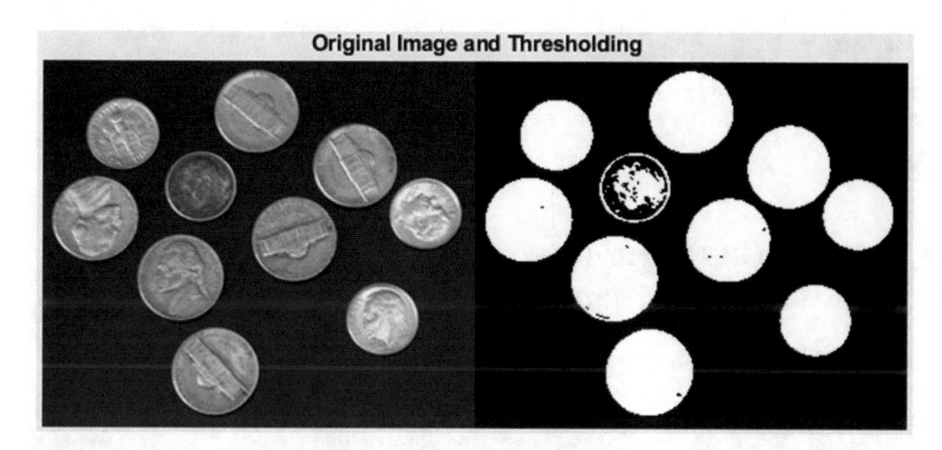

Figure 10-24. *Output of Example 10-19*

Double Thresholding

In double thresholding, two points are taken into account. Let us assume that threshold values are defined as T_1 and T_2. The pixel value at point (x,y) can be defined as follows:

$$P(x,y)=\begin{cases}\text{white, if the value at }(x,y)\text{ is between }T_1\text{ and }T_2\\ \text{black, if the value at }(x,y)\text{ is elsewhere}\end{cases} \quad (8)$$

Example 10-20. Write code to detect the borders of the `coin.png` file with $T_1 = 0.618$, and $T_1 = 0.382$ as thresholding values.

Solution 10-20. The following code can be used to accomplish the given task.

Example10p20.m

```
%Example10p20
%This code uses double thresholding
T1=0.618;T2=0.382;
im=mat2gray(imread('coins.png'));
[Row,Col]=size(im);
NewIm=zeros(Row,Col);
for i=1:Row
    for k=1:Col
        if (im(i,k)>T2) && (im(i,k)<T1)
            NewIm(i,k)=1;
```

```
        else
            NewIm(i,k)=0;
        end
    end
end
figure,imshowpair(im,mat2gray(NewIm),'montage');
title ('Original Image and Thresholding');
```

Figure 10-25 displays the output once the code has been run.

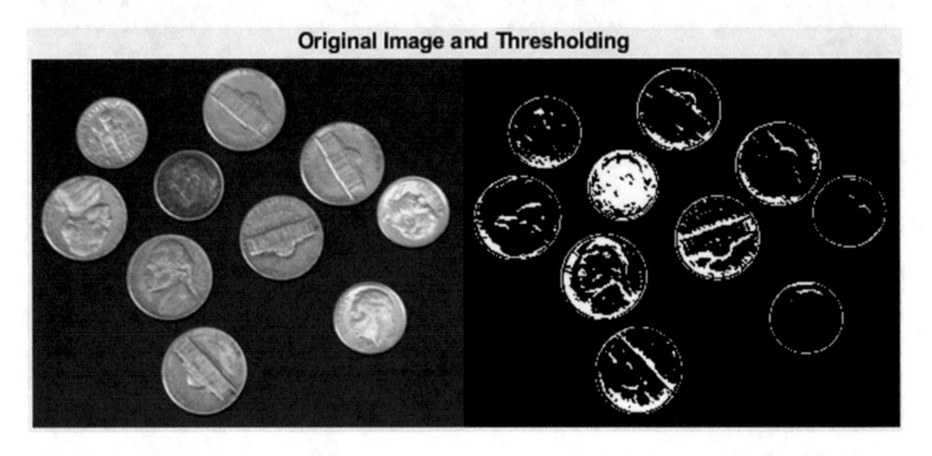

Figure 10-25. *Output of Example 10-20*

Edge Detection

Edge detection is defined as the detection of the boundaries of the objects within an image. It is one of the primary image processing techniques used for image segmentation and data extraction. There are different algorithms and methods used for this purpose in MATLAB. To find the edges in an image, the edge function is used. The methods used for this purpose include `sobel`, `prewitt`, `roberts`, `log`, `zerocross`, `canny`, and `approxcanny`. If any of these methods is not selected, the `sobel` method is used by default.

Many of the edge detection operators are constructed considering differentiation.

$$\frac{df}{dx} = \lim_{h \to 0}\left(\frac{f(x+h) - f(x)}{h}\right) \tag{9}$$

The definition of the derivative is given in Equation 9. This expression can be rewritten as

$$f(x+h)-f(x) \tag{10}$$

when $h = 1$. Actually, the smallest value of h is 1 if we consider the difference between consecutive pixel values horizontally.

Based on the expression $f(x + 1) - f(x - 1)$, horizontal and a vertical filters can be defined as $[-1, 0, 1]$, and $\begin{vmatrix} -1 \\ 0 \\ 1 \end{vmatrix}$. The magnitude of the gradient plays an important role in edge detection methods. For an image $f(x, y)$, the gradient and its magnitude are defined by

$$\left[\frac{\partial f}{\partial x}\frac{\partial f}{\partial y}\right], \sqrt{\left(\frac{\partial f}{\partial x}\right)^2+\left(\frac{\partial f}{\partial y}\right)^2}\ . \tag{11}$$

Example 10-21. Write a program that reads the image from the file `cowboys.png`. Then, use the methods `canny` and `prewitt` to find the edges within the image.

Solution 10-21. The following code can be used to accomplish the given task.

Example10p21.m

```
%Example10p21
%This code finds edges
Im = imread('cowboys.png');
Im2 = rgb2gray(Im);
CannyM = edge(Im2,'canny');
PrewittM = edge(Im2,'prewitt');
figure, imshow(CannyM);
title('Canny method');
figure, imshow(PrewittM);
title('Prewitt method');
```

Once the code is run, the outputs obtained are displayed in Figures 10-26 and 10-27.

Figure 10-26. *First output of Example10-21*

Figure 10-27. *Second output of Example 10-21*

Example 10-22: Two filters are defined as $P_x = \begin{bmatrix} -1 & 0 & 1 \\ -1 & 0 & 1 \\ -1 & 0 & 1 \end{bmatrix}$, and $P_y = \begin{bmatrix} -1 & -1 & -1 \\ 0 & 0 & 0 \\ 1 & 1 & 1 \end{bmatrix}$.

Using these two filters, find the edges of the `cowboys.png` image.

Solution 10-22. The following code can be used to accomplish the given task.

Example10p22.m

```
%Example10p22
%This code finds edges
I = imread('cowboys.png');
s = rgb2gray(I);
px=[-1,0,1;-1,0,1;-1,0,1];
py=px';
Sx=imfilter(s,px);
Sy=imfilter(s,py);
imshow(Sx),title('Applying Px');
figure,imshow(Sy),title('Applying Py');
```

Once the code is run, the output shown in Figures 10-28 and 10-29 is obtained.

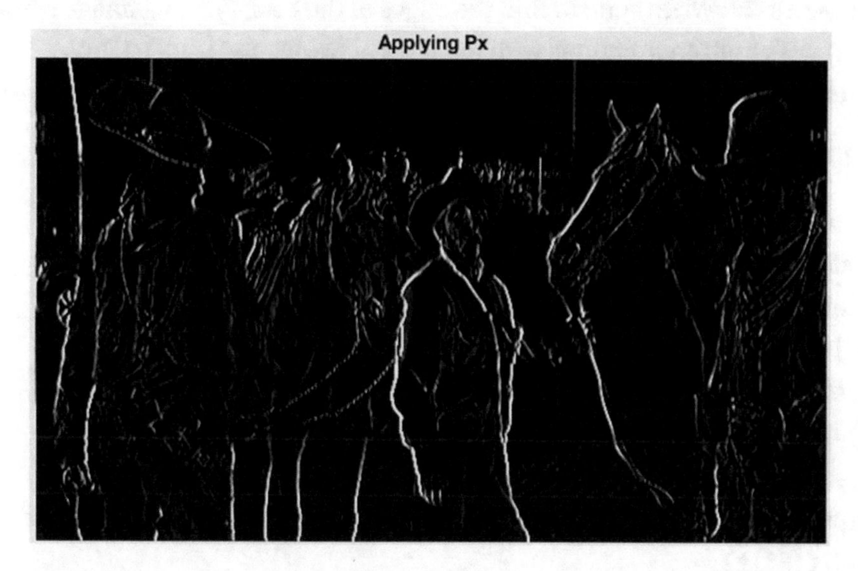

Figure 10-28. *First output of Example 10-22*

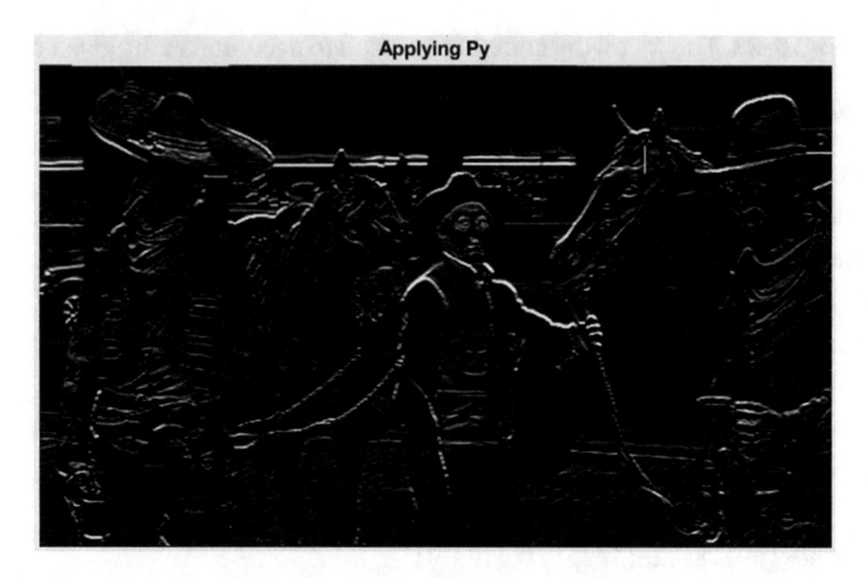

***Figure 10-29.** Second output of Example 10-22*

Example 10-23. Write code to find the edges of the `cowboys.png` image without using MATLAB's built-in functions with the filter of px=[0,-1,-1;1,0,-1;1,1,0].

Solution 10-23. The following code can be used to accomplish the given task.

Example10p23.m

```
%Example10p23
%This code finds edges without built-in functions
Im=double(rgb2gray(imread('cowboys.png')));
[Row,Col]=size(Im);
Im2=zeros(Row-1,Col-1);
Im2(1,1)=Im(1,1);
Filt=[0,-1,-1;1,0,-1;1,1,0];
Filt=flipud(Filt);
Filt=fliplr(Filt);
for i=2:size(Im, 1)-1
    for j=2:size(Im, 2)-1
        nghb_mtrx=Filt.*Im(i-1:i+1, j-1:j+1);
        avg_val=sum(nghb_mtrx(:));
        Im2(i, j)=avg_val;
    end
end
```

```
imshow(uint8(Im2));
title('Edge Detection');
```

In the preceding code, the convolutional operator is used. Once the code is run, the output shown in Figure 10-30 is obtained.

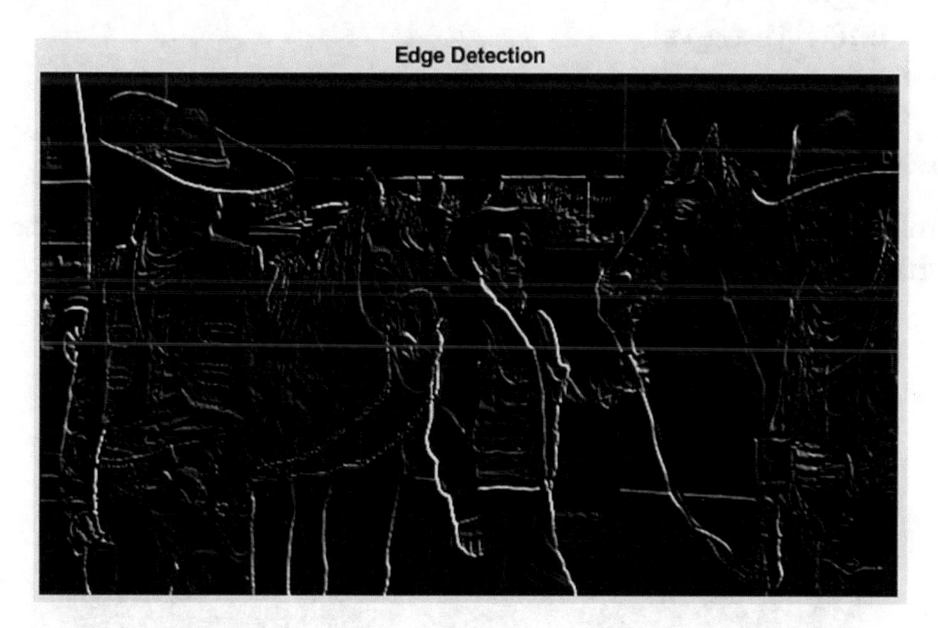

Figure 10-30. *Output of Example 10-23*

Example 10-24. Write code to find the edges of the `cowboys.png` image without using built-in functions with the filter of px=[0,1,0;1,-4,1;0,1,0], known as discrete Laplacian.

Solution 10-24. The following code can be used to accomplish the given task.

Example10p24.m

```
%Example10p24
%This code applies discrete Laplacian
Im=double(rgb2gray(imread('cowboys.png')));
[Row,Col]=size(Im);
Im2=zeros(Row-1,Col-1);
Im2(1,1)=Im(1,1);
Filt=[0,1,0;1,-4,1;0,1,0];
Filt=flipud(Filt);
```

```
Filt=fliplr(Filt);
for i=2:Row-1
    for j=2:Col-1
        nhbr_mtrx=Filt.*Im(i-1:i+1, j-1:j+1);
        avg_val=sum(nhbr_mtrx(:));
        Im2(i, j)=avg_val;
    end
end
imshow(uint8(Im2)); title('Discrete Laplacian');
```

In this code, convolution is used to obtain the output. Once the code is run, the output is shown in Figure 10-31.

Figure 10-31. *Output of Example 10-24*

Region-Based Methods

Generally, data clustering, region growing, region merging and splitting, and mean shift are the different methods used to apply region-based image segmentation. These methods attempt to find common features among the pixel values over a region on the image. One of the easiest methods to use is region growing. In this type, an initial seed point is picked, and then the difference between the seed point and the neighboring pixels is checked. If the difference is tolerable up to a certain point, or a threshold, then these pixels will be added to the seed pixel and the region will be grown. In this method, threshold value or criteria will be an important part of the algorithm.

Mathematical Morphology

Mathematical morphology is a branch of image processing for analysis and processing geometric structures such as extracting image components. Operations are performed for noise filtering, segmentation, shape simplifying, and enhancing object structure. Dilation and erosion are two important concepts in mathematical morphology. Dilation adds pixels to the boundaries of object and erosion removes pixels on the boundaries.

Assume that A and B are two sets having pixel values. Then dilation of A by B is defined as

$$A \oplus B = \{(x,y)+(u,v):(x,y) \in A,(u,v) \in B\}. \tag{12}$$

Erosion of A by B is defined as

$$A \ominus B = \{z : B_z \subseteq A. \tag{13}$$

All of the other operations are defined with a combination of dilation and erosion. Table 10-9 lists some of the most commonly used functions in morphology.

Table 10-9. *Functions Used in Morphology*

Function	Explanation
bwpack	Packs the entered binary image into `uint32`
bwunpack	Unpacks `uint32` binary image to binary image
imdilate	Dilates the grayscale, binary, or packed image
imerode	Erodes grayscale, binary, or packed image
imclose	Applies morphological closing on grayscale or binary image
imopen	Applies morphological opening on grayscale or binary image
imbothat	Applies morphological bottom-hat filtering to grayscale or binary image
translate	Translates a polyshape

Example 10-25. Write code that resizes the image of cowboys.png to 300 by 400. Then the code should use the strel function, which creates a cube structuring element with a width of 3 pixels. Then apply this to the resized image with erosion and dilation separately. The code should show the original image with the dilated one, and the original image with the eroded one in two different figures.

Solution 10-25. The following code can be used to accomplish the given task.

Example10p25.m

```
%Example10p25
%This code applies erosion and dilation
I= rgb2gray(imread('cowboys.png'));
BW=imresize(I,[300 400]);
se = strel('square',3);
BW1=imerode(BW,se);
BW2 = imdilate(BW,se);
figure,imshowpair(BW,BW1,'montage');
title('Original  with Erosion');
figure,imshowpair(BW,BW2,'montage');
title('Original  with Dilation');
```

Once the code is run, the outputs shown in Figure 10-32 and 10-33 are the result.

***Figure 10-32.** First output of Example 10-25*

Figure 10-33. *Second output of Example 10-25*

Problems

10.1. Write a program that reads the pixel values of a truecolor image in the file named `peppers.png`. Then, convert the format of the image to `.jpg` and save it as `New.jpg`.

10.2. Write a program that reads the image from the file `peppers.png`. Then, rotate the image 50 and 12 degrees.

10.3. Write a program that reads the images from the files `rice.png` and `cowboys.png`. Then, subtract these images from each other, and take the absolute value of the result. Finally, display these values as an image on the screen.

10.4. Write a program that reads the image named `rice.png`. Then, the code should generate two different images by using the pixel values of the read image. The pixel values of the first image should be the second power of the used image, and the pixel values of the second image must be obtained dividing the pixel values of the read image by three.

10.5. Write a program that reads the image named `peppers.png`. Then, R, G, B frequencies should be changed to B, R, G, and the code should display the final image.

10.6. Write code to adjust the pixel values of the image `peppers.png`. The code should saturate the upper 10% and lower 10% pixel values. The original and adjusted images should be plotted on a figure, including their histograms.

10.7. Write code to enhance the contrast of the `peppers.png` image using histogram equalization. The code should show the enhanced image, and difference image from the original to enhanced one in separate figures.

10.8. Write code to apply two different filters to the image `coins.png`. The filters are [1,1;1,1]∗1/4 , and [1,1,1,1,1;1,1,1,1,1;1,1,1,1,1; 1,1,1,1,1;1,1,1,1,1]∗1/25. The filtered images and the original image should be shown in separate figures.

10.9. Write code to apply the MATLAB filters of `gaussian`, `average`, and `motion` to `rice.png`. The filtered images should be shown in the same figure.

10.10. Write code that adds Gaussian white noise having zero mean and 0.01 variance to `peppers.png` image. Then use a 4-by-4 filter to remove the noise.

10.11. Write code that adds speckle multiplicative noise to the `peppers.png` image with noise density of 0.02. Then using an adaptive filter, remove the noise from the image. The code should show the original and filtered images in different figures.

10.12. Write code that reads the `peppers.png` image and splits the RGB image into its red, green, and blue channels. The code should show the size of the image and the first 10 elements from each channel on the screen.

10.13. Write code that multiplies every channel with different coefficients of the peppers.png image. The coefficients of red, green, and blue channels are 1/4, 3/4, and 1/2. After each channel is calculated with the given coefficients, channels should be combined to represent an RGB image. The code should show all of the updated images and the original image in one figure.

10.14. Write code to detect the borders of peppers.png with $T = 0.5$ as the threshold.

10.15. Write a program that reads the image from the file peppers.png. Then, use a Gaussian filter to find the edges of the image.

CHAPTER 11

Introduction to Sound Processing

It is possible to record, listen to, or manipulate sounds or audio files in MATLAB. This chapter introduces some basic features that are used with audio files, including using sine function as a sound. Using the Audio toolbox and other toolboxes, sophisticated illustrations can be achieved. In this chapter, though, the examples presented take into account that the reader might not have these toolboxes installed on MATLAB.

Sound propagates in the form of waves in the air. Sound is a continuous signal that can be digitalized as a discrete signal. Sound waves are recorded with a frequency that is the number of samples taken generally per second and measured in Hertz (Hz). Sounds with frequencies between 20 and 20,000 Hz comprise the hearing range for human beings.

To record and listen to a sound file, a microphone and a speaker should be installed on the system. To determine whether you have these devices, type the command `audiodevinfo` at the prompt.

```
> info = audiodevinfo
info =
     input: [1x2 struct]
    output: [1x2 struct]
>
```

In that code, the variable `info` is a construction having two fields. To visualize the content of `info` for the first data, you can type the following at the prompt.

```
> openvar info.input
```


I. Turk, *Practical MATLAB*, https://doi.org/10.1007/978-1-4842-5281-9_11

Then, the window shown in Figure 11-1 opens.

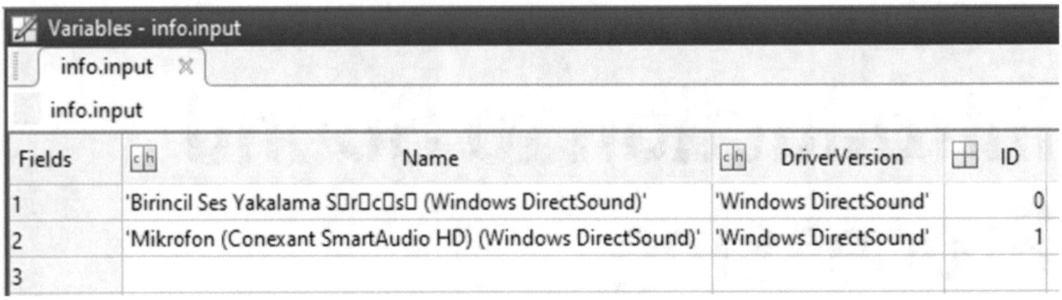

Variables - info.input

info.input

info.input

Fields	Name	DriverVersion	ID
1	'Birincil Ses Yakalama S□r□c□s□ (Windows DirectSound)'	'Windows DirectSound'	0
2	'Mikrofon (Conexant SmartAudio HD) (Windows DirectSound)'	'Windows DirectSound'	1
3			

Figure 11-1. *Microphone information*

If your operating system has a language other than English, you might see output similar to that shown in Figure 11-1. This display indicates that we have a microphone to record sound in MATLAB. To do that, we use the function `audiorecorder`. Using `audiorecorder`, we can create an 8,000 Hz, 8-bit, 1-channel audiorecorder file. It is also possible to change these default values. Possible numbers of bits include 8, 16, and 24. The number of channels should be either 1 (mono) or 2 (stereo).

Example 11-1. Write a program that loads the audio file `handel` to the workspace. Then, the program should play the sound.

Solution 11-1. The following code can be used to accomplish the given task.

Example11p1.m

```
%Example11p1
%This code plays hallelujah
load handel;
p = audioplayer(y, Fs);
play(p)
```

As shown, once the `handel` file is loaded, variables of *y* and *Fs* are loaded into the workspace. The variable *y* is the played signal, and *Fs* is the sample rate. The `audioplayer` function creates an `audioplayer` object for the signal *y*, and that object is assigned to the variable *p*. Then in the final row, using the command `play`, the *p* variable that has the sound is played.

To find information about a sound file, such as number of samples, number of channels, and so on, you can use the `audioinfo` as illustrated here.

```
> info = audioinfo('Adele.mp3')
info =
  struct with fields:
           Filename: 'C:\...\Adele.mp3'
  CompressionMethod: 'MP3'
        NumChannels: 2
         SampleRate: 44100
       TotalSamples: 12569472
           Duration: 285.0220
              Title: 'Someone Like You'
            Comment: []
             Artist: 'Adele'
            BitRate: 192
>
```

Example 11-2. Write a program that records your voice for 8 seconds with a sampling rate of 22,050 Hz, 16 bits with 2 channels. Then, the program should play the recorded sound and plot the graph of the recorded file.

Solution 11-2. The following code can be written to accomplish these tasks.

Example11p2.m

```
%Example11p2
% This records your voice for 8 seconds.
RecorderVar = audiorecorder(22050, 16, 2);
disp('Start speaking now')
recordblocking(RecorderVar, 8);
disp('End of Recording');
play(RecorderVar); % Playing
% Store data in double-precision array.
RecordData = getaudiodata(RecorderVar);
plot(RecordData); % plotting the sound
```

As shown in the code, the recorded sound is assigned to the variable RecorderVar. In the fifth row, the recordblocking function records the object variable for 8 seconds, and it does not return until the recording is finished. The play function plays the recorded sound. The getaudiodata function returns the recorded audio data as a double array. Once the data are plotted via the plot function, we obtain the graphic shown in Figure 11-2, which represents the recorded sound.

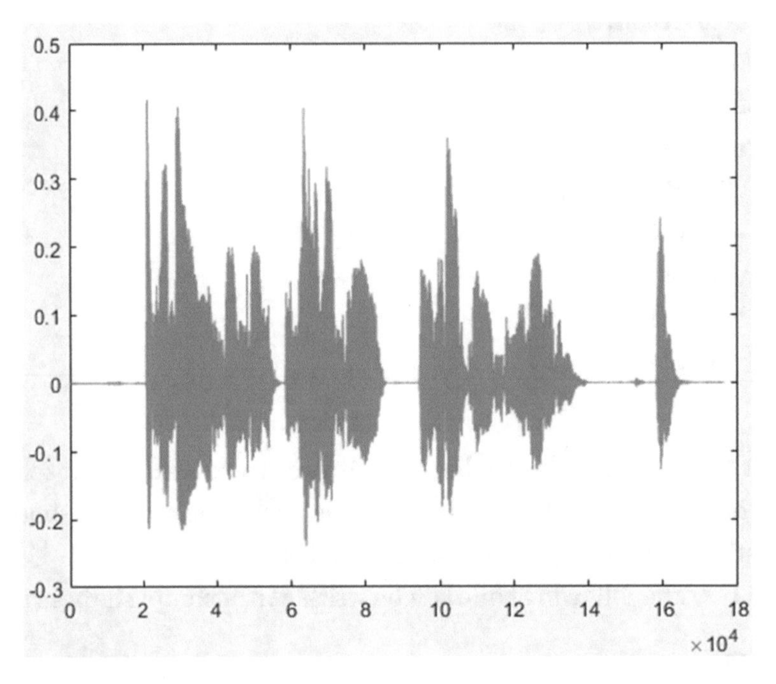

Figure 11-2. *Recorded sound*

Keep in mind that, when an individual runs the same code and speaks into the microphone, the output will be different from Figure 11-2 based on that person's voice. Plotting is also different from what was obtained with the plot function earlier due to the fact that raw audio data are used in Figure 11-2.

Example 11-3. Write a program that loads the handel audio file to the workspace. Then the program should save the file as handel.flac in the directory. The code should add some information to the file. In the comment field, the sound file should display "This is my first audio," in the title field "Hallelujah," and in the artist field "San Antonio." The bits per sample should be 24. The code should play the sound and display the information for the saved file.

Solution 11-3. The following code can be written to accomplish these tasks.

Example11p3.m

```
%Example11p3
% This code saves an audio file with info
load handel
filename = 'handel.flac';
audiowrite(filename,y,Fs,'BitsPerSample',24,...
'Comment','This is my first audio',...
'Title','Hallelujah','Artist','San Antonio');
sound(y,Fs);
clear y Fs
info = audioinfo(filename);
disp(info);
```

In the code shown, the sound function plays the vector y as sound. The function audiowrite writes the sound with the defined features to the directory. Once the program is run, the following output appears in the command window.

```
>Example11p3
            Filename: 'C:\...\handel.flac'
   CompressionMethod: 'FLAC'
         NumChannels: 1
          SampleRate: 8192
        TotalSamples: 73113
            Duration: 8.9249
               Title: 'Hallelujah'
             Comment: 'This is my first audio'
              Artist: 'San Antonio'
       BitsPerSample: 24
>
```

The played vector can have values between negative one and positive one. For other data types, the ranges are shown in Table 11-1 and the maximum number of channels is listed in Table 11-2.

***Table 11-1.** Range for y Depending on Its Data Type*

Data Type of y	Range of y
uint8	$0 \leq y \leq 255$
int16	$-32768 \leq y \leq +32767$
int32	$-2^{31} \leq y \leq 2^{31}-1$
single	$-1.0 \leq y \leq +1.0$
double	$-1.0 \leq y \leq +1.0$

***Table 11-2.** Maximum Number of Channels Depending on Data*

File Format	Maximum Channels
WAVE (`.wav`)	1,024
OGG (`.ogg`)	255
FLAC (`.flac`)	8
MPEG-4 AAC (`.m4a`, `.mp4`)	2

Example 11-4. Write code that loads the `handel` file to play. After playing for 3 seconds the file should stop playing and then resume. It should then play for 3 more seconds, and then stop.

Solution 11-4. The following code can be written to accomplish these tasks.

Example11p4.m

```
%Example11p4
% This code pauses and resumes playing
load handel;
player = audioplayer(y, Fs);
play(player);disp('playing is on');
pause(3);
pause(player);disp('playing is paused');
pause(3);
resume(player),disp('playing is resumed');
pause(3);
stop(player),disp('playing is off');
```

In this code, the play function starts the object player at the beginning. Then the pause function pauses the audio after 3 seconds of playing. The same pause function holds for 3 more seconds. The resume function resumes the playing and the program holds for another 3 seconds. Finally, the stop function stops playing the object. Once the code is run, the following output appears in the command window.

```
> Example11p4
playing is on
playing is paused
playing is resumed
playing is off
>
```

Example 11-5. Write code that loads handel file to the workspace. Then the code should reverse the audio and play it in the reversed order. The code also should print the first five data of the original audio, and the last five data of the reversed audio to check their accuracy.

Solution 11-5. The following code can be written to accomplish these tasks.

Example11p5.m

```
%Example11p5
% This code reverses the hallelujah
load handel;
L=length(y);
y2=zeros(L,1);
for i=1:L
    y2(L-i+1)=y(i);
end
sound(y2,Fs);
disp('First 5 Data of Original Handel');
disp(y(1:5));
disp('Last 5 Data of Reversed Handel');
disp(y2(end-4:end));
```

In this code, after loading handel, the length of y data is stored to the L variable. Then, a new variable y2 is initialized as having all zero elements. Then the values of y are stored in the y2 variable in reversed order. The code then plays the new data and displays the required data on the screen. Once the code is run, the following output appears in the command window.

```
> Example11p5
First 5 Data of Original Handel
         0
   -0.0062
   -0.0750
   -0.0312
    0.0062
Last 5 Data of Reversed Handel
    0.0062
   -0.0312
   -0.0750
   -0.0062
         0
>
```

Sine Wave as Sound

One can get a sound by using a sine wave defined as

$$y = A * \sin(2 * \pi * n * F_w / F_s) \tag{1}$$

where A is the amplitude of the wave, F_w is the frequency of the wave, F_s is the sample frequency, and $\boldsymbol{n}$ is the index of the sample. In some resources, the same equation is presented as

$$y = A * \sin(2 * \pi * \textit{frequency} * \textit{TimeVector}) \tag{2}$$

where $\text{frequency} = \text{sine frequency}$, and $\text{TimeVector} = \dfrac{\text{Total Number of Points}}{\text{Sample Frequency}}$.

Example 11-6. Write a program that plots the sine wave signal. The variables are amplitude = 32, total number of points = 1,000, sampling frequency = 1,000, and frequency of sine wave = 20. The code should play the sound of y according to Equation 1 or 2 and plot the signal.

Solution 11-6. The following code can be written to accomplish these tasks.

Example11p6.m

```
%Example11p6
% This code plays sine wave
A=32;N=1000;%Total number of points
Fs=1000;% Sampling frequency
frequency=20;%Frequency of sine wave
TimeVector=(0:N-1)/Fs;
y=A*sin(2*pi* frequency.*TimeVector);
sound(y);plot(TimeVector,y);
title('Sine Wave');ylabel('Amplitude');
xlabel('Time');
```

Once that code is run, a sound of the signal y is heard and the plot of the signal shown in Figure 11-3 is obtained.

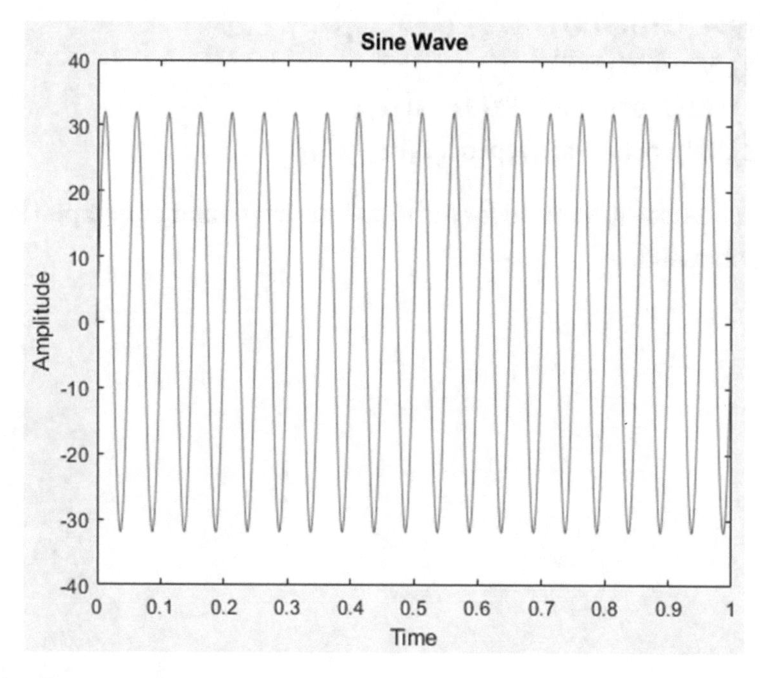

***Figure 11-3.** Sine wave plot*

Example 11-7. Write a program that plots the signal given by $ysin = A * (\sin(2 * \pi * freq * TimeVector)) + A * (\sin(4 * \pi * freq * TimeVector))$. The variables are amplitude = 1.5, total number of points = 1,500, sampling frequency = 1,000, and frequency of sine wave = 50. The code should take the `fft` of signal `ysin`, and should play and plot the signal.

Solution 11-7. The following code can be written to accomplish these tasks.

Example11p7.m

```
%Example11p7
% This plays sound
A=1.5;% Amplitude
N=1500;%Total number of points
Fs=1000;% Sampling frequency
freq=50;%Frequency of sine wave
TV=(0:N-1)/Fs;%time vector
ysin=A*sin(2*pi* freq.*TV)+...
    A*sin(4*pi* freq.*TV);
Y = fft(ysin);y = abs(Y/N);
P1=y(1:N/2+1);P1(2:end-1)=2*P1(2:end-1);
t=Fs*(0:(N/2))/N;sound(ysin);plot(t,P1,'r');
```

Once that code is run, a sound for the signal y is heard and output plot shown in Figure 11-4 is obtained.

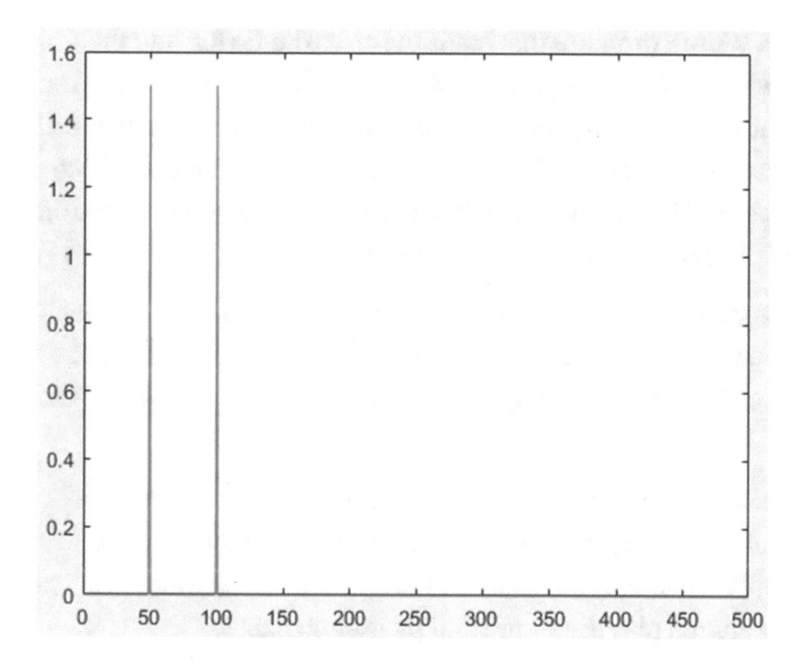

Figure 11-4. *Output of Example 11-7*

Two bars are shown in Figure 11-4. If we play with the coefficients of the sine function or amplitudes in the signal and change one of them, say that the amplitude of the first term is twice the amplitude of the second term in the signal, then we would get different values for the bars in the graphic.

Problems

11.1. Write a program that loads the audio files `chirp` and `handel` to the workspace. Then, the program should play both of the sounds.

11.2. Write a program that records your voice for 5 seconds with a sampling rate of 30,000 Hz, 16 bits with 2 channels. Then, the program should play the recorded sound back and plot a graph of the recorded file.

11.3. Write a program that loads the `chirp` audio file into the workspace. Then the program should save the file as `chirp.flac` in the directory. The code should add some information to the file. In the comment field, the sound file should have the value `Nice Voice`, and in the title field `Bird`. The code should play the sound and display the information for the saved file.

11.4. Write code that loads the audio file `chirp` to the workspace. The code should then separate the sound into two equal parts. The code also should play the second half first, and then the first half.

11.5. Write a program that plots the sine wave signal. The variables are amplitude = 10, total number of points = 5,000, sampling frequency = 8,000, and frequency of sine wave = 20. The code should play the sound and plot the signal.

CHAPTER 12

Applications with Graphical User Interfaces

A graphical user interface (GUI) can be created in three different ways. The first method is to create a GUI programmatically. In this instance, commands are usually written directly in the editor. The second way is to use the graphical user interface development environment (GUIDE). The third way is to create the GUI via the App Designer. In this chapter, we cover how to create GUIs in all of these cases, and illustrate how to program applications with them.

GUI Elements

Although in App Designer, there are more options available, the elements shown in Figure 12-1 can be used with GUIDE and programmatically.

I. Turk, *Practical MATLAB*, https://doi.org/10.1007/978-1-4842-5281-9_12

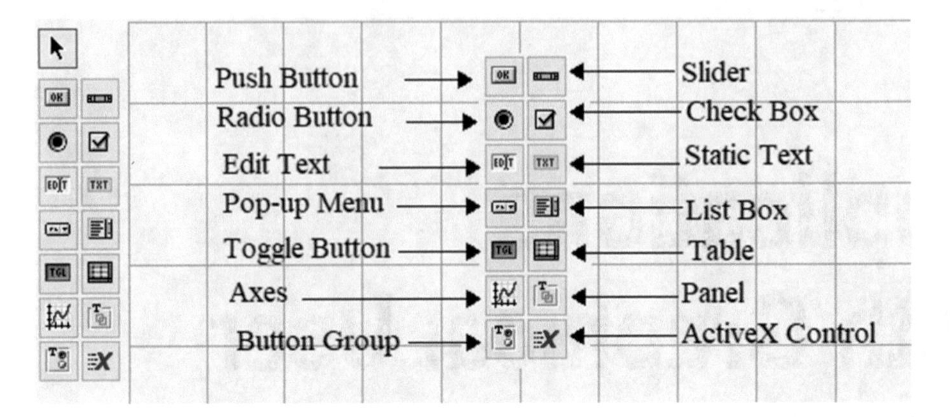

Figure 12-1. *GUI elements for GUIDE*

The buttons available for use are as follows:

- Push button
- Slider
- Radio button
- Check box
- Edit text
- Static text
- Pop-up menu
- List box
- Toggle button
- Table
- Axes
- Panel
- Button group
- ActiveX control

Using these buttons, it is possible to create neat and useful GUIs in MATLAB.

Creating GUIs Programmatically

Using this method, the buttons or elements can be directly defined and used in the editor.

Example 12-1: Write code that creates a GUI. The GUI will have two places to enter length and height values. Then a button should multiply these entered numbers and show the result in noneditable text output with a red background.

Solution 12-1. The following code can be used to accomplish the given task.

Example12p1.m

```
%Example12p1
% This code creates a GUI programmatically
close all;clear all; %[xpost,ypost,lngth,hght]
hFig=figure('NumberTitle','off','Menubar',...
'none','ToolBar','none','Position',...
[320 180 400 400],'NumberTitle','off',...
    'Name','My First GUI Programmatically');
uicontrol(hFig,'Style', 'text',...
'String', 'Enter Width: W','tag','htextW',...
'Position', [2 200 150 50],'FontSize',15);
uicontrol(hFig,'Style', 'text',...
'String', 'Enter Length: L',...
'Position', [2 300 150 50],'FontSize',15);
uicontrol(hFig,'Style','edit','String','0',...
   'position', [170 310 100 50],...
'FontSize',15,'tag','editL');
uicontrol(hFig,'Style','edit','String',...
'0','position', [170 210 100 50],...
 'FontSize',15,'tag','editW');
uicontrol(hFig,'Style', 'text',...
'String', ' 0 ','BackgroundColor','Red',...
'Position', [200 100 100 40],...
'FontSize',20,'tag','textResult');
uicontrol(hFig,'Style', 'pushbutton',...
'String', 'MULTIPLY L BY W',...
'Position', [20 100 150 50], 'Callback',...
{@Calculate},'FontSize',12);
```

```
function Calculate(hObject,event)
LL=findobj(0,'tag','editL');
L=str2double(get(LL,'String'));
WW=findobj(0,'tag','editW');
W=str2double(get(WW,'String'));
Result=findobj(0,'tag','textResult');
set(Result,'String',num2str(L*W));
end
```

Although this code looks very complicated, it is quite understandable if we look at each object separately. In the fourth row, `figure` is the main frame of our interface that holds information about the position of the GUI, including the name of the GUI. The expression `[320 180 400 400]` sets the position of the GUI as `[x-position, y-position, length, height]`. Then every set of rows starting with the `uicontrol` function creates an object in the GUI. There are two text objects to show texts, two editable objects to enter values, one button object to click on, and one more text object to show the result. Therefore, there are six rows starting with the `uicontrol` function that create these objects. One of them, which is the button object, has a callback function. When that button is clicked on the GUI, it calls a function called `Calculate`. Therefore, near the end of the code, the `Calculate` function is created to do the core part of the coding. If we run the code and enter the values shown on GUI, we obtain dialog box shown in Figure 12-2.

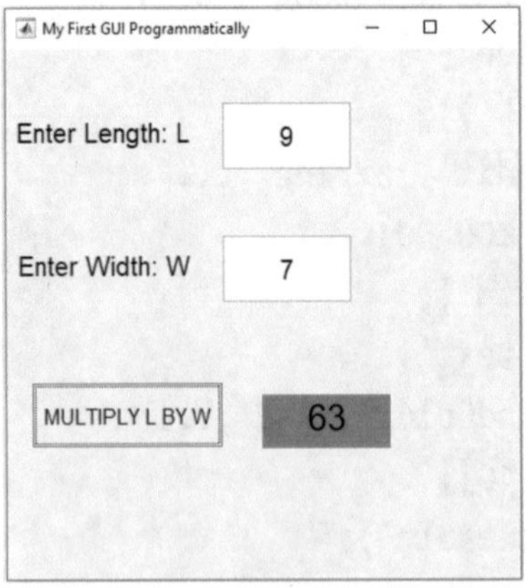

Figure 12-2. *Creating a GUI programmically*

Once we click the Multiply button, it calls the `Calculate` function and the entire code inside the function is run. The code should be able to get the entered values from the text boxes. This is performed inside the `Calculate` function. These objects are found by using their tag names. Therefore, in GUI programming, assigning tag names to the objects is extremely important. Using the `findobj function`, these objects are found and then their values are obtained using the `get` function. Using `str2double`, these values are converted to `double`. After the multiplication, the result is assigned to the text with the tag name is `textResult` via the `set` function.

Creating GUIs Using GUIDE

One can type `>guide` at the prompt to view the GUIDE. After opening, it asks whether the user wants to open a new GUI or an existing GUI. In this scenario, you should open a new one and create the following objects by dragging and dropping them on the canvas.

Example 12-2. Write a program to code the GUI as shown in Figure 12-3.

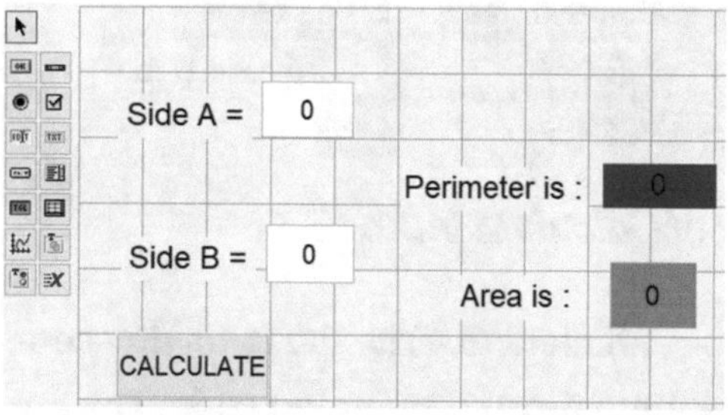

Figure 12-3. *Creating a GUI using GUIDE*

Solution 12-2. We need to drag the components from the left and drop them on the right side as shown in Figure 12-3. Then, if you double-click the icons, or right-click the icon and select Property Inspector, you can change the font size of the strings, values, string names, tags, and some other features by choosing from the available options within the inspector. Tag names are important in programming the GUI because tags are referred to whenever the components are contacted, as explained before. The tag name for the Side A text box is `editA`, for the Side B text box the tag name is `editB`, for the Perimeter text box it is `textshow`, and for the Area text box it is `editson`.

After creating the GUI in Figure 12-3 and saving it as `Example12p2.fig`, a function named `Example12p2.m` is automatically generated by MATLAB.

If you run the function Example12p2 from the command window, or click the green button in Figure 12-3, then the created function starts working. If you just click the Calculate button on the GUI, though, nothing will happen. At minimum, we need a function that will request the values of A and B and then, display the area and perimeter after clicking the Calculate button.

The Calculate button is tagged as pushbutton1 in the inspector for the button (Figure 12-4).

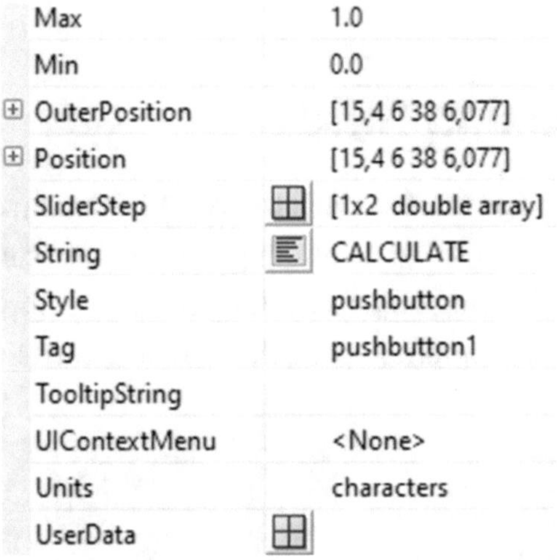

Property	Value
Max	1.0
Min	0.0
⊞ OuterPosition	[15,4 6 38 6,077]
⊞ Position	[15,4 6 38 6,077]
SliderStep	[1x2 double array]
String	CALCULATE
Style	pushbutton
Tag	pushbutton1
TooltipString	
UIContextMenu	<None>
Units	characters
UserData	

Figure 12-4. *Inspector of Calculate button*

In our case, that will be sufficient to write the following code under the pushbutton1_Callback function in the automatically generated code. The function names might differ depending on the tag names you assign to the elements.

```
AA=get(handles.editA,'String');
A=str2double(AA);
BB=get(handles.editB,'String');
B=str2double(BB);
Area = A * B;
Perimeter= 2*(A+B);
set(handles.textshow,'String',Perimeter);
set(handles.editson,'String',num2str(Area));
```

In the first line of the preceding code, the string value is obtained. In the second line, the value is converted into `double` for length `A`. The same procedure is repeated for length `B` in the third and fourth lines. Area and perimeter values are calculated by using the formulas in the fifth and sixth lines. Finally, these values are assigned to the places that were tagged as `textshow` and `editson`.

The `pushbutton1_Callback` function should look like the Figure 12-5.

```
% --- Executes on button press in pushbutton1.
function pushbutton1_Callback(hObject, eventdata, handles)
% hObject    handle to pushbutton1 (see GCBO)
% eventdata  reserved - to be defined in a future version
% handles    structure with handles and user data (see GUI
AA=get(handles.editA,'String');
A=str2double(AA);
BB=get(handles.editB,'String');
B=str2double(BB);
Area = A * B;
Perimeter= 2*(A+B);
set(handles.textshow,'String',Perimeter);
set(handles.editson,'String',num2str(Area));
```

***Figure 12-5.** Additional function written within the generated code*

Once the code is executed, the GUI shown in Figure 12-6 is the result. If we try the GUI with the values A = 10 and B = 61, and click Calculate, we get the result shown.

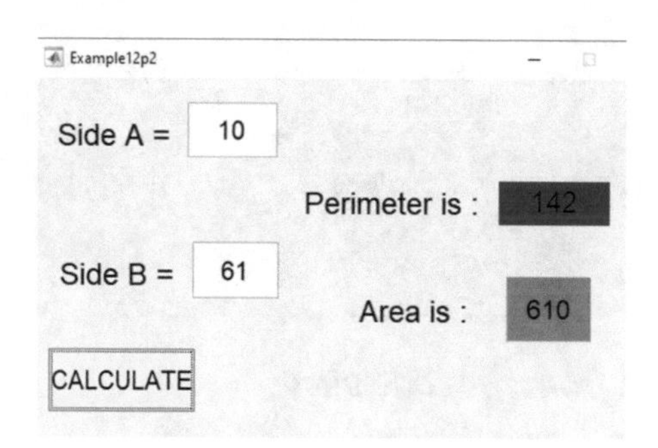

***Figure 12-6.** Output of the GUI for the given values*

Creating GUIs with App Designer

To create any application, we need to open the App Designer first. On the MATLAB toolbar, one can click the New button and then the App option will be available to select. After selecting App, the interface of the app will appear.

On the left, one can select the GUI elements. The objects can be dragged and dropped to the center of the canvas to be created. The easiest part is the creation of objects. The important part is writing callback functions here, too. In App Designer, MATLAB gives you permission to write on the white places in the coding part. As an example, one can just write codes on the white spaces seen in Figure 12-7. The rest of the coding is automatically generated depending on the design of objects.

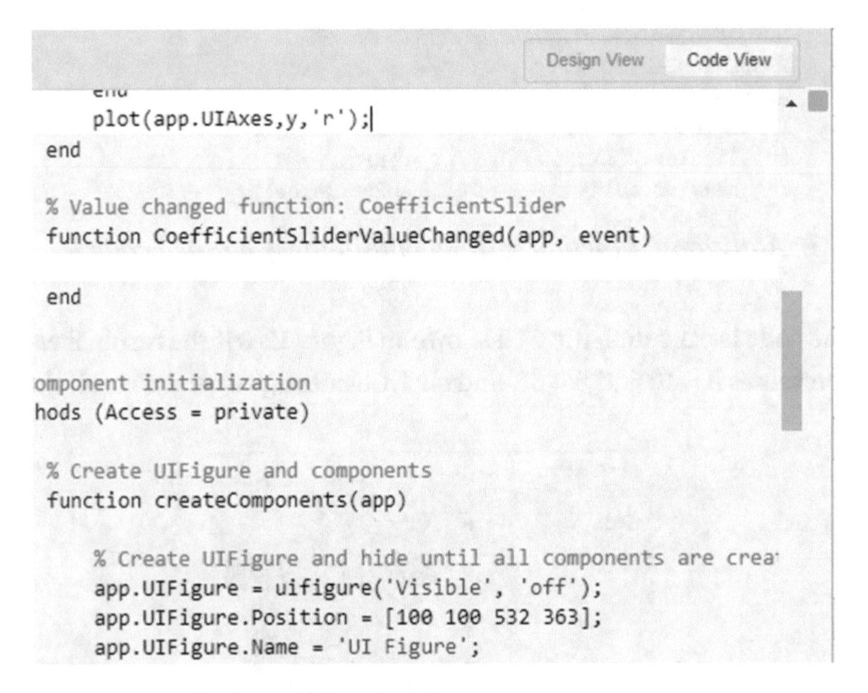

Figure 12-7. *An outlook from the code view*

Example 12-3. Write a program with App Designer to plot the graphics of trigonometric functions of sine, cosine, and tangent between $-\pi$ and π. These functions should be selected from a `listbox`. There should also be a slider to multiply the selected function. Once the Plot button is clicked, the code should plot the trigonometric function that is multiplied by the coefficient of the selected value of the slider.

Solution 12-3. After designing the GUI, we need to create a callback function. Figure 12-8 shows code used as a callback function.

```
% Button pushed function: PLOTButton
function PLOTButtonPushed(app, event)
    Coeff=double(get(app.CoefficientSlider,'Value'));
    Option=get(app.FunctionsListBox,'Value');
    t=-pi:0.01:pi;
     switch Option
        case 'Sin'
            y=Coeff.*sin(t);
            disp(y);disp('sin');
        case 'Cos'
            y=Coeff.*cos(t);
            disp(y);disp('cos');
        case 'Tan'
            y= Coeff.*tan(t);
            disp(y);disp('tan');
    end
    plot(app.UIAxes,y,'r');
end
```

Figure 12-8. *Code from* `Myapp1.mlapp`

Figure 12-9 shows the output once the code is run.

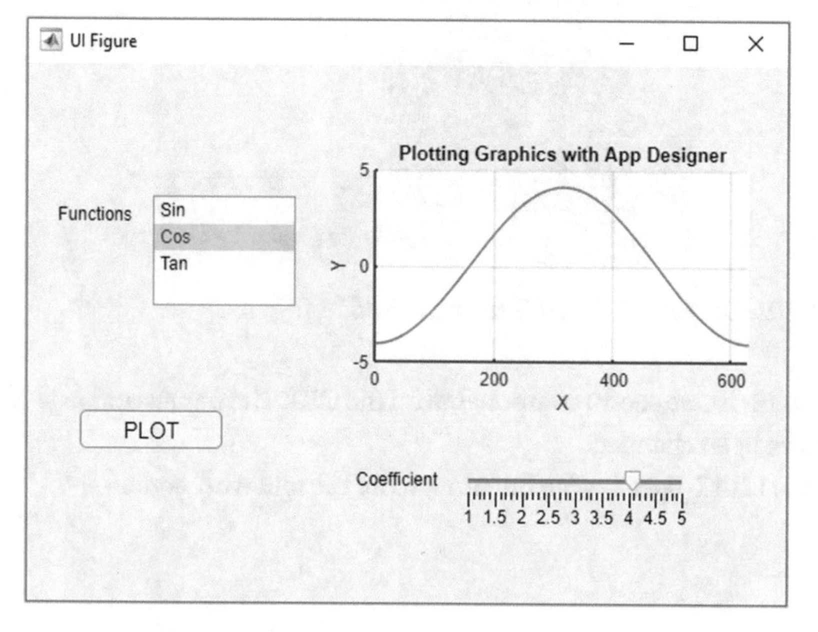

Figure 12-9. *Obtained output*

What is different from GUIDE is that `app.` should be added as prefix to each tag, and the names of the files have an `.mlapp` suffix instead of `.m`. The rest of the coding is almost the same as coding with GUIDE, although depending on the elements or components library, App Designer has more available options.

Creating Applications

Many applications can be created using GUIs in MATLAB. In this section, a few of them are illustrated in the examples. These applications are created by GUIDE.

Example 12-4. Create a GUI that has four options for plotting the graph of a trigonometric function such as sine, cosine, tangent, and cotangent of x, where x is between 0 and 2π. The function should be multiplied by a constant having a value between 0 and 100. The value of the constant should be controlled using a slider.

Solution 12-4. Figure 12-10 shows the outline of the GUI.

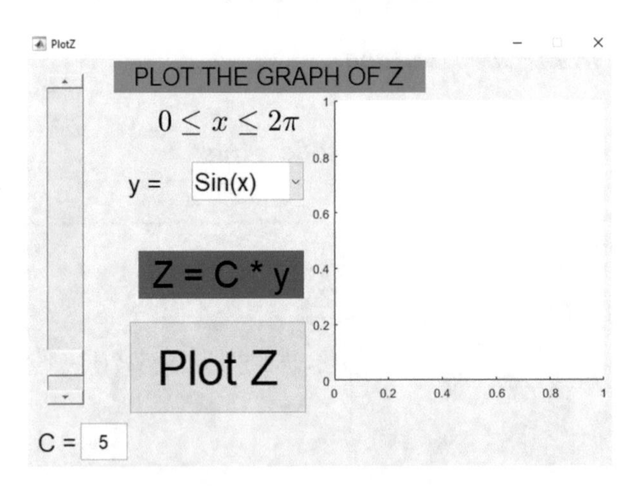

***Figure 12-10.** Output of the GUI named PlotZ*

In Figure 12-10, we see 10 components on the GUI. The names and tags of the elements have been changed.

Under the `PlotZ_OpeningFcn` function, write the following code.

```
axis(handles.axesSymbol,'off');
ht = text(1,0.5,'$0\leq x\leq 2\pi$',...
    'HorizontalAlignment','Right',...
    'Interpreter','latex','FontSize',25);
set(ht,'Parent',handles.axesSymbol);
Slide    = get(handles.sliderC, 'Value');
Slider = num2str(Slide);
set(handles.editC, 'String',Slider);
```

Under the sliderC_Callback function, write the following code.

```
Slide    = get(handles.sliderC, 'Value');
Slider = num2str(Slide);
set(handles.editC, 'String',Slider);
```

Under the editC_Callback function, write the following code.

```
Slide    = get(handles.editC, 'String');
Slider = str2double(Slide);
set(handles.sliderC, 'Value',Slider);
```

Finally, under the pushbuttonZ_Callback function, write the following code.

```
ChoicE = get(handles.popupmenuY,'Value');
CC=get(handles.editC,'String');
C=str2double(CC);
x=0:0.1:2*pi;
switch ChoicE
    case 1 % sin(x)
        z=C*sin(x);
        plot(z);
    case 2 %cos(x)
        z=C*cos(x);
        plot(z);
    case 3 %tan(x)
        z=C*tan(x);
        plot(z);
    case 4 %cot(x)
```

```
        z=C*cot(x);
        plot(z);
end
grid on
```

Once we run the code, we obtain the GUI shown in Figure 12-10.

Example 12-5. Create a GUI for a restaurant to calculate customers' bills. The logo and the name of the restaurant should be on the GUI as well. Within the GUI, there should be panels for soups, desserts, and other available meals. Customers can pick only one kind of soup, but they can pick as much as they desire of anything else.

Solution 12-5. Figure 12-11 displays the outline of the GUI.

Figure 12-11. *Output of the GUI named* `Restaurant.m`

After designing the GUI as shown, and saving it as `Restaurant.fig`, the file `Restaurant.m` is automatically generated. Then, the codes given next are entered for the corresponding functions.

As a first step, the logo of the restaurant should be uploaded to the GUI. Hence, under the `Restaurant_OpeningFcn` function, write the following code.

```
axis(handles.axes1,'off');
myImage = imread('Restaurant.jpg');
axes(handles.axes1);
imshow(myImage);
```

In the preceding code, the logo is saved as `Restaurant.jpg` within the same directory as the files `Restaurant.m` and `Restaurant.fig`.

Next, we need to write code to calculate the bills for the selected meals. Under the `pushbuttonprice_Callback` function, write the following code.

```
global Amount
Soup = get(handles.uibuttongroupsoup,'SelectedObject');
PickedSoup = get(Soup,'Tag');
switch PickedSoup
    case 'radiobuttontomato'
        Soupprice = 5;
    case 'radiobuttonchicken'
        Soupprice = 7;
    case 'radiobuttonlentil'
        Soupprice = 8;
    otherwise
        msgbox('You did not pick ANYTHING');
end

Baklava = get(handles.radiobuttonbaklava,'Value');
Cake    = get(handles.radiobuttoncake,'Value');
RicePudding= get(handles.radiobuttonricepudding,'Value');
MixedFruit=get(handles.radiobuttonmixedfruit,'Value');
Shish       = get(handles.checkboxshishkebab,'Value');
Bursa     = get(handles.checkboxbursaiskender,'Value');
Wrapper     = get(handles.checkboxwrapper,'Value');
Rice        = get(handles.checkboxrice,'Value');
Salad       = get(handles.checkboxsalad,'Value');
Compote     = get(handles.checkboxcompote,'Value');
Soda        = get(handles.checkboxsoda,'Value');
Coffee      = get(handles.checkboxcoffee,'Value');
HotTea      = get(handles.checkboxhottee,'Value');
```

```
Amount = Soupprice+Baklava*5+Cake*5+RicePudding*5+...
    MixedFruit*7+Shish*20+Bursa*22+Wrapper*15+...
   Rice*10+Salad*10+Compote*7+Soda*3+Coffee*3+HotTea*2;
set(handles.textsonuc,'String',Amount);
```

Under the pushbuttontotal_Callback function, write the following piece of code.

```
global Amount
if Amount<1
    msgbox('You did not pick ANYTHING');
else
    Tipp    = get(handles.edittip, 'String');
    Tippp = str2double(Tipp);
    TotalPrice = Amount + Tippp;
    set(handles.texttotal,'String',TotalPrice);
end
```

After running the program, the GUI is ready to work.

Example 12-6. Build a GUI that emulates a calculator.

Solution 12-6. Figure 12-12 shows the output of the code.

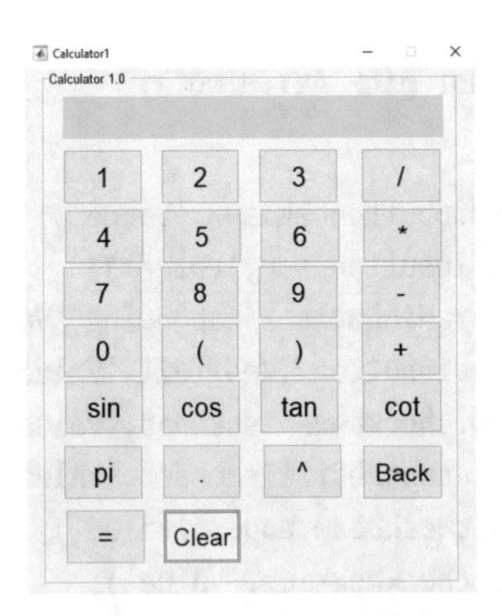

Figure 12-12. *Outline of the GUI named* `Calculator1.m`

It is necessary to write the callback functions for each button on the GUI. Each button is tagged with its name. For example, 2 is tagged as Two, + is tagged as Add, and so on.

Under the One_Callback function, write the following code.

```
OldVal=get(handles.textresult,'String');
NewVal='1';
TextString = strcat(OldVal,NewVal);
set(handles.textresult,'String',TextString);
```

Under the Two_Callback function, write the following code.

```
OldVal=get(handles.textresult,'String');
NewVal='2';
TextString = strcat(OldVal,NewVal);
set(handles.textresult,'String',TextString);
```

We repeat the same thing for buttons 2, 3, 4, 5, 6, 7, 8, 9, and 0. For division, nter the following code under the Divide_Callback function in our case.

```
OldVal=get(handles.textresult,'String');
NewVal='/';
TextString = strcat(OldVal,NewVal);
set(handles.textresult,'String',TextString);
```

Similarly, repeat this procedure for the operations ∗, -, + and the symbols (.) and (,) as well as π for their corresponding buttons. For the sin button, enter the following code under the Sine_Callback function.

```
OldVal=get(handles.textresult,'String');
NewVal='sin((pi/180)*';
TextString = strcat(OldVal,NewVal);
set(handles.textresult,'String',TextString);
```

The angle is converted to degrees from radians in this code. That's why we need to write `(pi/180)*` within the second line of the code. Similar code must be written for the cos, tan, and cot buttons. For the Clear button, enter the following code under the Equal_Callback function.

```
EvValue   = get(handles.textresult,'String');
EvalValue = eval(EvValue);
set(handles.textresult,'String',EvalValue);
```

In that code, the function eval evaluates the expression EvValue. Under the Back_Callback function, enter the following code.

```
textString = get(handles.textresult,'String');
if(strcmp(textString,'0')==1)
   set(handles.textresult,'String','0') ;
else
    Val=char(textString);
    L=length(textString);
    textString=Val(1:L-1);
set(handles.textresult,'String',textString)
end
```

After running the code, the GUI is ready to work.

Example 12-7. Construct a GUI for the guess game.

Solution 12-7. Figure 12-13 shows the output of the code.

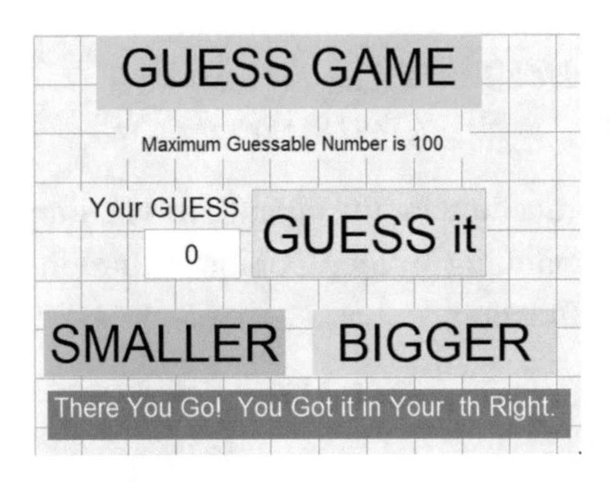

***Figure 12-13.** General view of the GUI*

The program is required to guess a number whenever it is started over. Therefore, we need to write code under the opening function of the GUI, which will be called GuessGame_OpeningFcn, as shown in the following.

```
global Picked_Number Counter
Picked_Number=randi(100,1);
Counter=1; %initializing counter
```

The first line, which includes the global variables, should be placed right after the function line. The next two lines can be entered at the end of the function.

Under the pushbuttonGuess_Callback function in our case, enter the following code.

```
global Picked_Number Counter
GuessVal=get(handles.editWriteGuess,'String');
Guess=str2double(GuessVal);

if Guess==Picked_Number
    fprintf('You got it in your %d th right\n',Counter);
    set(handles.textBigger,'visible','off');
    set(handles.textSmaller,'visible','off');
    Message1 = 'There You Go! You got it in your ';
    Message2 = 'th right!';
    FinalMes = [Message1, num2str(Counter), Message2];
    set(handles.textYouGot,'String',FinalMes);
    set(handles.textYouGot,'visible','on');
    return;
elseif Guess>Picked_Number
    set(handles.textYouGot,'visible','off');
    set(handles.textBigger,'visible','off');
    set(handles.textSmaller,'visible','on');
elseif Guess<Picked_Number
    set(handles.textYouGot,'visible','off');
    set(handles.textSmaller,'visible','off');
    set(handles.textBigger,'visible','on');
end
Counter=Counter+1;
```

After running the code, the GUI is ready to work.

Example 12-8. Create a music player GUI. The music should be loaded by a button on the workspace.

Solution 12-8. Figure 12-14 displays the output of the code.

***Figure 12-14.** General view of the created music player*

The picture should be loaded in the GUI. Therefore, the following code should be put at the end of the `MusicPlayer_OpeningFcn` function as shown here.

```
axis(handles.axes1,'off');
myImage = imread('Figure_12_14.png');
axes(handles.axes1);
imshow(myImage);
```

Similarly, the following code should be put at the end of the `Play_Callback` function as shown in the following.

```
global filename player
[y,Fs] = audioread(filename);
player = audioplayer(y,Fs);
play(player);disp('playing is on');
```

Similarly, the following code should be put at the end of the `Pause_Callback` function.

```
global player
pause(player);
```

The following code should be put at the end of the `Resume_Callback` function.

```
global player
resume(player);
```

The following code should be put at the end of the `Stop_Callback` function.

```
global player
stop(player);
```

The following code should be put at the end of the `Load_Callback` function.

```
global filename  pathname
[filename, pathname] = ...
    uigetfile({'*.mp3';'*.wav'});
```

After running the code, the GUI is ready to work.

Example 12-9. Write code that deals one card from a deck of cards.

Solution 12-9. Figure 12-15 shows the output of the code.

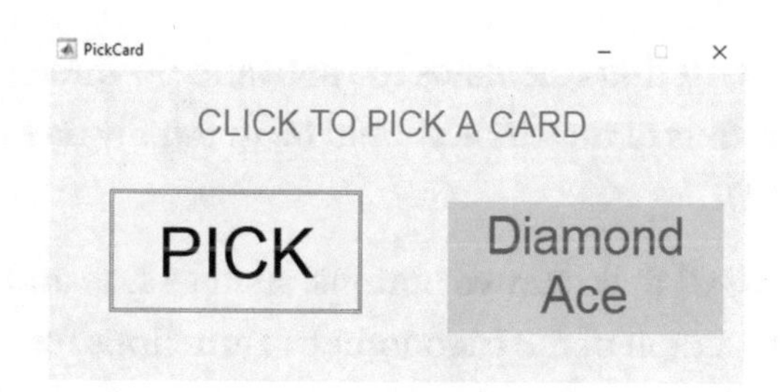

Figure 12-15. *General view of the GUI*

After creating the GUI components, the code is saved as `PickCard`. The `pick_callback` function will be similar to the following.

```
% --- Executes on button press in pick.
function pick_Callback(hObject, eventdata, handles)
% hObject    handle to pick (see GCBO)
% eventdata  reserved - to be defined in a future %version of MATLAB
% handles    structure with handles and user data  %(see GUIDATA)
Suit = {'Club', 'Spade', 'Heart', 'Diamond'};
Numbers={'Ace','2','3','4','5','6','7','8',...
    '9','10','Jack','Queen','King'};
Su=randi(4,1);Nu=randi(13,1);
Card=[cellstr(Suit(Su)),cellstr(Numbers(Nu))];
%Tipp    = get(handles.edittip, 'String');
set(handles.textR,'String',Card);
```

After running the code, the GUI is ready to work.

Many applications can be created using GUIs. Most of the time, GUIDE is used as to create these applications, but App Designer has been used more recently and more components have been added to the canvas as well. There is a `Migrate` tool to convert an application created in GUIDE into App Designer files. It can easily be downloaded from the Mathworks web site.

Problems

12.1. Create a GUI that calculates the area and perimeter of a circle. The radius of the circle should be manually entered through the GUI.

12.2. Create a GUI that has two options, such as sine and cosine, for plotting the graph of the trigonometric functions for the values of x, where x is between 0 and π. After the user clicks the Plot button, the GUI should plot the graph.

12.3. Create a GUI that performs as a basic calculator with the four basic math operations (addition, subtraction, multiplication, and division).

12.4. Create a GUI that deals 13 cards to a player. The code should show all of these cards on the GUI.

12.5. Create a GUI for a restaurant showing five different meals. A customer can give more than one order for the same meal. The GUI should calculate the total amount the customers spent.

12.6. Create a GUI that plays an MP4 video. Stop and Pause buttons should be on the GUI as well.

APPENDIX A

References

1) Allen, Linda J. S. *An Introduction to Mathematical Biology*. Upper Saddle River, NJ: Pearson, 2006.
2) McAndrew, Alasdair. *A Computational Introduction to Digital Image Processing*. Boca Raton, FL: CRC, 2016.
3) Mandal, Mrinal, and Amir Asif. *Continuous and Discrete Time Signals and Systems*. London: Cambridge University Press, 2007.
4) Edwards, C. Henry, and David E. Penney. *Differential Equations and Boundary Value Problems: Computing and Modeling*. Upper Saddle River, NJ: Prentice Hall, 2003.
5) Gonzalez, Rafael C., and Richard E. Woods. *Digital Image Processing*. Boston: Pearson, 2017.
6) Hahn, Brian H., and Daniel T. Valentine. *Essential MATLAB for Engineers and Scientists*. Cambridge, MA: Academic Press.
7) Lial, Margaret L., Raymond N. Greenwell, and Nathan P. Ritchey. *Finite Mathematics*. Boston: Pearson, 2011.
8) `https://github.com/jueseph/Tutorial-GrowthCurves-MATLAB`
9) Hall, Barry G., H. Acar, A. Nandipati, and M. Barlow. Growth Rates Made Easy. *Molecular Biology and Evolution, 31* (1), 2014.
10) Cristianini, Nello, and Matthew W. Hahn. *Introduction to Computational Genomics, A Case Studies Approach*. London: Cambridge University Press, 2006.
11) Lent, Graig S. *Learning to Program with MATLAB: Building GUI Tools*. Hoboken, NJ: Wiley, 2013.
12) Moore, Holly. *MATLAB for Engineers*. Boston: Pearson, 2017.

I. Turk, *Practical MATLAB*, https://doi.org/10.1007/978-1-4842-5281-9

13) Attaway, Stormy. *MATLAB: A Practical Introduction to Programming and Problem Solving.* Oxford, UK: Butterworth-Heinemann, 2018.

14) Turk, Irfan. *MATLAB Programming for Beginners and Professionals.* Scotts Valley, CA: CreateSpace, 2018.

15) Apostol, Tom M. *Mathematical Analysis.* Englewood Cliffs, NJ: Pearson, 1974.

16) Banon, Gerald Jean Francis, Junior Barerra, and Ulisses de Mendonça Braga-Neto. *Mathematical Morphology and its Applications to Signal and Image Processing: Proceedings of the 8th International Symposium on Mathematical Morphology.* Rio, de Janeiro, Brazil: MCT/INPE, 2007.

17) Ding, Ding. Modeling and Simulation of Highway Traffic Using a Cellular Automaton Approach, UUDM Project Report, 2011.

18) `https://www.mathworks.com/help/`

19) `https://www.mathworks.com/help/images/linear-filtering.html`

20) Cheney, E. Ward, and David R. Kinciad. *Numerical Mathematics and Computing.* Pacific Grove, CA: Brooks/Cole, 2007.

21) Kinciad, David, and Ward Cheney. *Numerical Analysis: Mathematics of Scientific Computing.* Providence, RI: American Mathematical Society, 2002.

22) Zucker, Steven W. Region Growing: Childhood and Adolescence. *Computer Vision, Graphics, and lmage Processing, 5* (3), 1976.

23) `http://www.stat.rice.edu/~mathbio/stat300/`

24) Nowak, M., and R. May. *Virus Dynamics: Mathematical Principles of Immunology and Virology.* Oxford ,UK: Oxford University Press, 2001.

25) `http://www.yildiz.edu.tr/~naydin/na_I2B.htm`

26) `https://web.ma.utexas.edu/users/davis/375/popecol/lec10/lotka.html`

Index

I. Turk, *Practical MATLAB*, https://doi.org/10.1007/978-1-4842-5281-9

H

I, J, K

S

T

U

V

W, X, Y, Z

Printed in the United States
By Bookmasters